CONTINENT OF CURIOSITIES

Collecting curiosities was a popular pastime for wealthy, educated eighteenth-century European gentlemen, and few creatures aroused more curiosity than those that arrived from Australia. But collections demand to be organised, and the process of classification reveals patterns to life that cannot be ignored. From a leisurely occupation, the science of biology was born. *Cabinets de curiosités* expanded to become national museums, with specimens from Australia playing an integral role in all kinds of biological debates. Australian museums now foster their own research and continue to provide major and sometimes unexpected perspectives on international developments in many areas of science, from anthropology to space exploration.

Continent of Curiosities follows the thread of individual natural history stories inspired by specimens and scientists of one of Australia's oldest museums, Museum Victoria. Together, these stories weave an eclectic path through the history of biological science from an Australian perspective, with insights into the people and places which influence the way we see and understand the natural world around us.

Dr Danielle Clode is a lecturer in zoology at the University of Melbourne with an interest in the history of Australian biology Her previous book, *Killers in Eden*, is now a major ABC TV documentary. She is currently working on an account of early French naturalists in Australia.

CONTINENT *of* CURIOSITIES

A Journey Through
Australian Natural History

DANIELLE CLODE

CAMBRIDGE
UNIVERSITY PRESS

CAMBRIDGE UNIVERSITY PRESS
Cambridge, New York, Melbourne, Madrid, Cape Town, Singapore, São Paulo

Cambridge University Press
477 Williamstown Road, Port Melbourne, VIC 3207, Australia

Published in the United States of America by Cambridge University Press, New York

www.cambridge.org
Information on this title: www.cambridge.org/9780521866200

First published 2006

Printed in China through Bookbuilders

A catalogue record for this publication is available from the British Library

National Library of Australia Cataloguing in Publication data
    Clode, Danielle.
    Continent of curiosities: a journey through Australian
    natural history.
    Bibliography
    Includes index.
    ISBN-13 978-0-521-86620-0 hardback
    ISBN-10 0-521-86620-0 hardback
    1. Museum Victoria – History 2. Naturalists – Victoria – Melbourne. 3. Natural history –
    Australia. 4. Natural history museums – Victoria – Melbourne. I. Title.
069.07509945

ISBN-13 978-0-521-86620-0
ISBN-10 0-521-86620-0

MUSEUMVICTORIA

CONTENTS

FOREWORD

THIS BOOK EXPLORES the magic of museums. We are all
captivated, I think, by the awe and fascination that Danielle Clode feels
when she enters our great collection palaces. What lies in their basements?
What stories stir among the artefacts of nature and of history that sit
together in museum cabinets?

That great British hunter of natural curiosities, Alfred Russel Wallace,
who makes an appearance in these pages, was considered 'a conjurer' by
the Malayan islanders he interrogated in the 1850s. He was collecting
shells, insects, birds and animals, hunching over them with intense
concentration, drying and preserving specimens, reverently packing them
and taking them away in boxes. Islanders watched him at work in the
jungle and concluded that he must have been a shaman! And in his way
he was. Science has generally fought to distinguish itself from magic, but
sometimes it has enjoyed the confusion. Wallace's mind, fresh from the
intellectual ferment of industrial Britain, was spinning a theory with these
curiosities. His specimens were captured and suspended in time so that
they could voyage across the earth. They were destined for a museum,
where they would tell a story. But what story? What was the connection
between this thing in his hands, so recently stilled, and the ideas that
excited him and his society?

Danielle Clode's book pursues that question through the basements
and back rooms of a great museum. Museum Victoria, which she visited as
a child, has since opened other doors to her, those doors that lead beyond
the bright, hallowed halls into 'a parallel world'. She describes the mystery
and enchantment of its hidden corridors and their denizens. Best of all, she
takes us by the hand and shows us how particular, collected objects become
endowed with meaning. Museums are sensuous places, particularly in that
semi-ordered state behind the scenes. Sights, sounds and smells surround
one. There is unexpected beauty and horror to be found.

As is appropriate in a collection of scientific detective stories, Clode
offers us a journey back through time – in four stages. First, through

hundreds of years, then thousands, then millions and finally billions, to the formation of planets and the origins of life. But each chapter begins in the present, with the surviving evidence that propels us on our historical and scientific quest.

Clode was supported in this work by Museum Victoria's Thomas Ramsay Science and Humanities Fellowship. This is a wonderful scheme, funded by a bequest, which fosters research and writing across both the sciences and the humanities, that great divide in our intellectual culture. The Museum, by offering such a fellowship, rightly sees itself as an institution that is uniquely placed to bridge these ways of seeing. The collections themselves, in all their materiality and specificity, demand a holistic eye.

Over a decade ago, I was privileged to be a writer in this same museum, also supported by the Ramsay Fellowship. I worked in a corner of the museum overlooked by a stuffed lion in permanent but outmoded rage. At afternoon tea I could talk with Aboriginal people who had travelled far to consult and sometimes claim collections they regarded as theirs. Every day I walked past the gorillas that the foundation director of the museum had imported to Australia in the 1860s to disprove evolution. Museums spill over with stories that are quite likely to contradict one another; they are collections of intellectual fashions as well as of objects; they are institutions that struggle to impose linearity on the labyrinth and bring order to the attic – but fail gloriously. There are no better places to be promiscuous among the evidence, interdisciplinary, even undisciplined, and to be continually reminded of the contingencies of interpretation. The ballast of physical heritage in the basements and back rooms of museums will, hopefully, give these treasured institutions some stability as the storms of short-term management sweep across them.

One of my favourite metaphors for historical research is that of dredging a pond. If the world is a deep pond and we live on the surface, swimming so that we can also breathe, then the historian's job is to dredge the pond, keeping it healthy by continually disturbing the water and its contents. The surface is a busy but – by definition – superficial place, and there is limited room in the limelight at any one time. Things that were once given favoured attention on the surface later sink into the murky depths, forgotten. Historians dredge, continually dredge. Diving can be scary and hazardous, and you can't afford to stay down too long. But the quest is compelling: to remember, remind, discover, bringing to the surface half-familiar shapes, disturbing the superficial present with evidence from the depths.

Museums, galleries and libraries are those depths. They are full of organic matter, they steam with fetid fertility, they glow with a thousand auras; their true purpose utterly resists rationalisation. They are the greenhouses of our emotional and intellectual gardens. Here are the seeds of ideas and insights; here are the nutrients and warmth and light to help nurture the imagination; here is the deep, rich, smelly earthiness of composted life from which new life grows. I've just described what sounds like a conservator's nightmare, but I think you know what I mean: our collecting institutions are the source of the stories that sustain and disturb us, they are the memory-palaces of our culture.

There is another reason to celebrate this book. It reminds us of the essential link between research and collections in a period when museums increasingly have come to privilege the manager and the designer over the research curator. The stories in this book remind us that objects and ideas are symbiotic, interdependent, and that research at the moment of accession is especially important. There is a danger that public collections can neutralise and diminish the power of objects; some things should never be in collections and some should be let out from time to time. It's important, then, not just to collect the objects alone; we need to collect the stories that hover like a halo about them. It is the paraphernalia that surrounds a relic or specimen – meanings, ideas, functions, feelings, relationships – that makes it come alive again.

So we must ensure that institutional collecting means more than preserving a physical entity. It must also mean conserving an original context. And it necessarily entails creating something new – introducing that object to a world of scholarship and articulating its relationships with the present. The moment for research must be grasped. Collecting institutions must be research institutions.

So, let's follow Danielle Clode now as she takes us on a journey – into the museum's collections and back through time, travelling on the adrenalin of ideas.

TOM GRIFFITHS
February 2006

ACKNOWLEDGEMENTS

WITHOUT THE INSPIRATION of the curators at the Museum Victoria, the idea for this book would never have germinated. Without the support of the Thomas Ramsay Science and Humanities Fellowship Committee, the book would never have been written. Without the gentle persistence of Robin Hirst and the unwavering efforts of Jenny Darling, Jacinta di Mace and Donica Bettanin, the book might never have been published. For all their encouragement and support in helping this project reach fruition, I am most grateful. Museum Victoria has also provided generous financial support to assist with publication of the book.

Writing a book that covers a diversity of specialties is always a daunting experience and I have relied on the guidance and expertise of many people both from Museum Victoria and elsewhere. Their patience in guiding me through their collections and their research, and their efforts to keep my subsequent musings accurate, have been invaluable, but I take sole responsibility where I have wavered from the path. In particular I would like to thank the following from Museum Victoria (both past and present): Bill Birch, Les Christidis, John Coventry, Tom Darragh, Joan Dixon, Richard Gillespie, Martin Gomon, Forbes Hawkins, Dermot Henry, John Long, Wayne Longmore, Richard Marchant, Rory O'Brien, Tim O'Hara, Gary Poore, Tom Rich, Dianne Riggs, Gaye Sculthorpe, Ron Vanderwal, Ken Walker, Robin Wilson and Alan Yen.

Over the last ten years, my association with the Department of Zoology at the University of Melbourne has enabled me to stay in touch with my own scientific discipline and given me access to an invaluable array of support and expertise. The assistance of the following people has been much appreciated: Rachel Allan, Terry Beattie, Graeme Coulson, Sharon Downes, Mark Elgar, Kath Handyside, June Hook, Garry Jolley-Rogers, David Macmillan, Angus Martin, David Paul, Marilyn Renfree, Laila Sadler, Simon Ward, Barbara Wells, and David Young. Also from Melbourne University, I would like to thank Peter Attiwill, Mark Burgman and Terry Walsh for their assistance with matters botanical and Rod Holmes for gentle guidance on matters historical.

Many other people have read chapters and offered their thoughts, encouragement or criticism and I am grateful for every one of these contributions. In particular I would like to thank Barry Butcher from Deakin University for his insights into the Ape Case, Tom Griffiths and Libby Robin for their thoughts on the Forests of Fire and Terry Walsh from Melbourne Water for his knowledge of Melbourne's water history. The assistance of Patricia Vickers-Rich at Monash University has cropped up in a surprising array of topics, from dinosaur brains and avian fossils to pre-Cambrian life forms. The patient professionalism of Mark Adams and Charles Hussey from the Museum of Natural History, London, in providing data on the acquisition of Wallace bird specimens was greatly appreciated.

Paul Davies and the late Stephen J. Gould were both kind enough to offer their feedback on the chapters on Mars and Trigonia respectively and I am most grateful for their patience, time and encouragement. The encouragement of my undergraduate lecturers Paul Corcoran and the late Frank Dalziel from the University of Adelaide has also been an ongoing source of sustenance to me over many years. Paul's encouraging words to a struggling first-year kept me at university a lot longer than I would ever have imagined, while Frank's ability to justify lectures on dinosaurs in a psychology course proved that anything is possible with a bit of creative thinking.

There have been times when it seemed that the easiest part of this book was writing it. I would like to thank Ian Galloway and Robin Hirst for their ongoing support over the years in my various guises at the museum. Garry Warner and John Kean opened my eyes to a new way of seeing and interpreting science. Aranka McDonald and Jannine Allan provided beautiful illustrations. The assistance of Melanie Raymond, Ingrid Unger, Marija Bacic and John Kean was essential in saving me from drowning in a sea of unsourced images. The efforts and encouragement of Kim Armitage, Sally Chick, Margot Jones, Susan Keogh and Janet Mackenzie at Cambridge University Press kept me going when I could see no end in sight.

Finally, I must thank Jenny Lee and Carolyn Rasmussen for being both mentors and role models. My parents Anne O'Brien and John Clode have patiently read chapters of various books and covered their bemusement at my choice of occupation most tactfully. But most of all I must thank my children Lauren and Rachel – for giving me the best excuse in the world to stay home and write books instead of getting a proper job; and my husband Michael Nicholls – for encouraging me to write but preventing me from becoming obsessed.

DANIELLE CLODE
January 2006

Part 1

VISIONS FROM
THE OLD WORLD:
THE LAST 500 YEARS

Years ago

500 — 1500 – Pinzò presents the King of Spain with an opossum skin

1536 – Portuguese priest describes a scaly-tailed rat with a pouch from the Spice Islands

1578 – Publication of *Speculum Orbis Terrae* with early maps of Australia

1600 – Reverend Topsell draws a wild beast named Su

400

1659 – Huygens identifies polar ice caps on Mars

1699 – Tyson dissects a chimpanzee

300

1705 – De Bruijn dines on 'Aru rabbits' in Java

1735 – Linnaeus names the opossum 'two-wombed'

1770 – Banks has a 'kanguru' shot near Cooktown

1802 – Péron finds a trigonia on a beach in Tasmania

200

1832 – Darwin collects a sparrow in Uruguay

1853 – National Museum of Victoria founded

1859 – Wallace watches the birds in Bali

1864 – Meteorite strike in Orgueil heard across France

1881 – World's tallest tree cut down in Gippsland

100

1912 – Wegener proposes the theory of continental drift

1919 – Chisholm disturbs a barking spider eating a chicken

1939 – 10 per cent of Victoria burns in massive fires

1966 – An 'extinct' Mountain Pygmy Possum found in a ski lodge

1983 – Thomson Dam built in Victoria

1987 – Skull imprint of *Leaellynasaura amicagraphica* discovered

2000 – Melbourne Museum opens in Carlton Gardens

NOW

Nat Size

W. W. Froggatt del.   Troedel & C.º Print   R. Wendel lith. Melbourne

1 CURIOUS
COLLECTIONS

Great Pampa-Finch (*Embernagra platensis*), collected in 1832
in Maldanado, Uruguay, by Charles Darwin.
Ornithology Collection, Musuem of Victoria.

WHEN I WAS A CHILD, the exhibition halls of our natural history museum seemed endless, with row upon row, cabinet after cabinet, of rocks, shells, stuffed and pickled animals, strange bones, enormous eggs and bizarre agglutinations. Little did I know, as I toured the cool, dark halls of our local museum, tugging on my grandmother's arm, that these public displays barely even scratched the surface of museum collections. I could not have imagined that only a fraction of the vast collections housed in museums are ever displayed. I did not realise that the small, succinct explanatory notes are but a minuscule synthesis of the vast body of scientific literature born of museum collections.

It was many years before I gained a closer insight into the hidden nature of museums. By then I had turned my childhood passion for nature into a career as a biologist. But increasing familiarity has done little to diminish the awe and fascination which museums are capable of inspiring – not so much their exhibition halls, but the scientists and collections behind them. Beyond the lofty spaces of the museum exhibition spaces buzzing with crowds of excited children, behind the impenetrable transparency of the glass display cases, beside the brightly lit panels of interesting facts there exists a parallel world. This is a world of window-less, winding corridors – a world of darkness, death and some very strange smells. This is a quiet world of endless collecting, gradual sifting and patient preparation. Here lie unacknowledged treasures – skulls, skins, fossils and feathers – their true value rarely revealed by their spidery labels. Yet each of these individual objects exists within a rich and vibrant tapestry of knowledge and understanding about the world we live in. Even the most inconsequential specimen can play a part in a story which spreads from local personalities and events to theories that have changed the way we see the world – from historical curiosities to contemporary environmental crises.

This is the world I'd like to explore in this book, and I hope the stories related will provide an insight into the secret back rooms of both museums and biology. There are, of course, limitless stories I could have drawn on. Choosing which ones to focus on and which to leave out has been a difficult task, but the ones that are left I hope illustrate both the breadth and depth of the influence that museum specimens, collections and scientists have in the broader field of biological science. The eleven specimens which inspired the following stories link to issues as diverse as the European discovery of Australia, indigenous knowledge, old-growth forests, water use, palaeontology, brain physiology, evolution, creationism, biogeography, conservation, climate change, exploration and discovery. Some stories have a local and contemporary focus while others stretch

forwards into the future and back to the beginning of life itself. Others extend far beyond Australian shores to Europe, America, Asia and beyond, into outer space.

Small wonder then, that if my feet ached after circumnavigating the exhibition spaces, it was nothing compared to my aching head after a day behind the scenes among the 15 million or so specimens in Museum Victoria's natural history collection with their curators. Every specimen has a story – where and how it was collected, who by and what it meant at the time. Every object is part of a bigger narrative – how the species evolved, interacted with other species, how they are distributed, have expanded, contracted and disappeared.

One such object lay in my hand on my first visit to the ornithology collection. It was just a small brown bird, a soft ball of fluff hardly distinguishable to the untrained eye from a sparrow. It was, in fact, a Great Pampa-Finch (*Embernagra platensis*), one of the 160-odd species of seed-eating Emberizine finches found on the grassy savannahs of South America. What made this particular specimen intriguing was that it had been collected in 1832 in Uruguay, at the time when the *Beagle* sailed along the South American coast on its way to circumnavigate the world via the Galapagos Islands, Australia and New Zealand. The reason this particular bird is handled with such reverential awe by biologists is that the tag on its leg identifies the name of its collector – Charles Darwin (1809–1882). This sense of connection with the past is something only collections like those in museums can provide.

Brush-tailed Phascogale (*Phascogale tapoatafa*), one of Australia's marsupial carnivores (*Dasyuridae*), whose ferocity belies its small size.

J. Allan.

THE NEED TO CLASSIFY

Give any small child a container of buttons and you will see that a collection demands classification. Humans have an intrinsic desire to seek patterns in their world. The ability to formulate conceptual categories allows us to understand complex and seemingly ever-changing phenomena. But classification has some startling consequences when applied to collections of plants and animals. The patterns of nature are not only aesthetic, but are also a silent testimony of the history and origins of life on Earth.

Imagine a collection of interesting mammals. From Australia – a Brush-tailed Phascogale (*Phascogale tapoatafa*), a Sugar Glider (*Petaurus breviceps*) and a Greater Bilby (*Macrotis lagotis*). From North America – an Antelope Jack Rabbit (*Lepus alleni*), a Western Harvest Mouse (*Reithrodontomys megalotis*) and a Northern Flying Squirrel (*Glaucomys sabrinus*). And from South America – a Mara (*Dolichotis patagonum*), a Woolly Opossum (*Caluromys lanatus*) and a Vampire Bat (*Desmondus rotundas*).

Imagine that they are classified according to how they live. Gliding or flying, air-borne mammals might be placed in one drawer, while the arboreal mammals which spend their lives entirely in vegetation might be placed together in another. Terrestrial mammals, which typically live and rest on the ground, might logically be placed in a third drawer. But this classification by habitat also correlates with a classification by body structure. All of the gliding/flying mammals have membranes which allow them to 'fly'. The arboreal creatures all have adaptations for climbing – strong claws, flexible arm and leg sockets, prehensile or balancing tails. And all of the terrestrial animals have elongated legs and balance on their toes (digitigrade feet) allowing faster locomotion. Organising the mammal collection by habitat reveals a lesson in adaptation.

ADAPTIVE GROUPING	GEOGRAPHIC GROUPING		
	Australian	South American	North American
Flying	*Sugar Glider*	Vampire Bat	Squirrel Glider
Tree-dwelling	*Phascogale*	*Opossum*	Harvest Mouse
Ground-dwelling	*Greater Bilby*	Mara	Jack Rabbit

Reading the table horizontally shows the adaptive grouping; the vertical reading shows the geographic grouping. The physiological grouping into marsupial or placental is shown with the marsupials in italics.

Alternatively, the mammal collection could be organised by fundamental body structures, such as the reproductive system. This system of classification reveals a clear dichotomy – marsupials in one drawer and placental mammals in another drawer. This classification cuts across the adaptive features of the previous system and redistributes the specimens into completely different groups. But again, another unexpected pattern emerges – most of the marsupials are Australian species (with the exception of a South American opossum) while most of the placental specimens are American. In fact, the marsupials seem to be a feature of the older, more isolated land masses of the southern hemisphere (Australia and South America), while the placental animals appear to be a feature of the large interconnected land masses of Africa, Eurasia and, until relatively recently, North America. Biological patterns start to suggest geological histories of the land masses they come from – the beginnings of biogeographic theory.

The fastidious curator might, at this stage, decide to reorganise the collection entirely along geographic lines, with group membership determined by the continent of origin. Bearing in mind the findings of our earlier classifications, the curator might be surprised to notice that each continental grouping seems to contain just one member of each of the ecological groupings we began with. The Australian group contains one gliding possum, one arboreal phascogale and one terrestrial bilby. The North American group also contains a glider (squirrel), an arboreal harvest mouse and a terrestrial rabbit. The South American group contains a bat, an arboreal opossum, and a terrestrial mara. Although the animals are quite unrelated to one another, each region has developed similar animals to fill particular niches.

What began as a simple attempt to decide which specimens belong in which drawers of a cabinet has revealed several things: that geography underlies differences in basic physiological history; that adaptive distinctions from basic physiology are driven by the demands of different habitats; and that fundamentally different stock on different land masses appear to converge through the pressures of similar habitat and life strategies to look superficially similar. The patterns of nature are the fundamental observations of all evolutionary biology and these patterns led geneticist Theodore Dobzhansky to declare that 'Nothing in biology makes sense except in the light of evolution.'

There is something ineffably significant about seeing with your own eyes a bird collected by Darwin during the formative years of his evolutionary theory – or the skull of a giant 12-metre goanna (*Megalania prisea*) which stalked the Australian landscape within human history – or the silky skin of a Lesser Bilby (*Macrotis leucura*) which barely survived European activities in Australia long enough to be identified by science. Museum specimens link us physically, culturally and intellectually to our past. They are objects endowed with a special resonance and authenticity that technology and interpretation can augment and enhance, but never replace. Perhaps this is why spaces crammed with the paraphernalia of the past can be so engaging compared to the sometimes over-designed and interpreted spaces of many modern museums. Just as the detritus and heirlooms of daily life beloved of local history museums reveal the world of our grandparents, natural history specimens reveal the natural world as it is, as it once was, how we used to see it, and how it might be in the future.

We are all capable of astonishing leaps of imagination and creativity to link these unfamiliar objects into our known world. Sometimes we are fortunate to travel in the company of a knowledgeable guide – a grandmother who remembers Aunty Beryl buying one of those orange juicers for 2s. 6d. or an uncle whose knowledge of steam engines is unparalleled. I count myself as fortunate indeed to have passed briefly through the back rooms of Museum Victoria in the company of many

A display case of objects from Museum Victoria.

Part 1 VISIONS FROM THE OLD WORLD

collection curators. An offhand comment about a specimen is enough to elicit a whirlwind of connections and coincidences. A pretty shell might spiral into a discussion of geographic variation associated with water depth and ocean temperatures, or offer a view from the first large-scale oceanic survey in Australian waters, or an insight into the life of an extraordinary, but almost unknown, collector. In the hands of their curators, the mundane, the rare and the dazzling, all become keys to other worlds.

It is the self-appointed mission of scientists to navigate through unfamiliar objects, facts, discoveries or observations and attempt to pull them into some kind of coherent pattern. For the museum scientist, the objects in their collections are tangible facts – the physical manifestation of the physicist's data, the chemist's reaction, the mathematician's formula or the zoologist's observation. The objects in a natural history collection, and how they are interpreted, offer a unique insight into the scientific mind and the scientific process.

Today, biological research is conducted in a vast range of institutions, from universities to government research organisations to private industry. But museums, the founding institutions of biological science, remain important contributors to scientific debates. Carefully tending and adding to their priceless and invaluable collections built up over centuries, museum curators provide the historical backbone to scientific research, in which it is all too easy to forget work conducted fifty years ago in favour of that conducted within the last five years. Ongoing developments in evolutionary biology, in particular, are often founded on the fundamental, but time-consuming, work of taxonomists and museum collections.

The type of research conducted in a modern museum differs greatly from the research that first led to their foundation in previous centuries. When Melbourne's museum first opened, many of Victoria's birds and mammals were still poorly understood and the museum provided the only opportunity for the public to compare these species with those from overseas. Today, there is less demand for collecting new vertebrates, although discoveries of reptiles, fish and invertebrates continue to be made. The original collections are now important indicators of the past diversity and abundance of many threatened species and form a baseline for conservation research. While the less charismatic creatures of the world have always found a home in museums, the need to document and understand them is even more urgent today. One of the most active areas of museum-based research is the expansion of marine, freshwater and terrestrial invertebrate collections in collaboration with conservation programs and environmental studies.

VICTORIA'S MUSEUM

Frederick McCoy
(1823–1899), Director of
the National Museum of
Victoria, 1858–1899.

The institution now known as Museum Victoria began its life as the Museum of Natural and Economic Geology in 1854, one of the first museums to be founded in Australia. In 1856 the professor of natural science at the University of Melbourne, Frederick McCoy, moved the collection to the university and in 1858 was appointed director of the National Museum of Victoria, a position he was to hold for over forty years.

McCoy employed an aggressive strategy for increasing the museum's collections. His third occupational hat, as palaeontologist to the Victorian Geological Survey, provided a steady stream of fossil material to the museum's collections. Alternately petitioning the government for money, purchasing specimens in advance, wheedling exchanges and pestering suppliers, McCoy built one of the finest natural history collections in the southern hemisphere. McCoy was joined by a taxidermist, John Leadbeater, in 1859 (followed by William Kershaw in 1864), but the bulk of the work of identifying, cataloguing, displaying and managing the collections fell to McCoy.

In 1887 Walter Baldwin Spencer (1860–1929) joined the University of Melbourne staff as the first professor of biology. As one of only a few professional biologists in Victoria, Spencer's

The interior of the
National Museum
of Victoria.
State Library of Victoria.

early work also contributed to the museum, most notably during the Horn Expedition to central Australia in 1894. This expedition, whose collections now reside in the museum, was pivotal in shifting Spencer's original interest in zoology towards the ethnography of the indigenous inhabitants of Australia. Spencer took on responsibility for the museum following the death of McCoy in early 1899, moving it from the university to a temporary home in the State Library. During the thirty years Spencer reigned as director, he laid the foundations for the museum's collections of indigenous Australian artefacts and cultural records as well as material from across the Pacific.

The staff at the museum had increased in number during this time, with a specialist palaeontologist, entomologist and taxidermists joining the institution. William Kershaw, the preparator of specimens was succeeded by his son James Kershaw (1866–1946), who progressed through the ranks to become a curator, then director following Spencer's retirement. Ornithologists and conchologists later joined the staff. Another preparator to move through the curatorial ranks to become director was Charles Brazenor in the late 1950s. The museum now employed specialist curators of mammals, birds, insects, molluscs, fossils, minerals and anthropology. These scientists, and their successors over the next fifty years, not only managed their collections and conducted their research, but also played important roles in public education by developing exhibitions and acting as a reference point for innumerable public enquiries. Many took on the responsibility of management: the majority of the museum's directors have been scientists of one kind or another.

In 1983 the National Museum of Victoria was amalgamated with the Science Museum of Victoria to form the Museum of Victoria. In 1998 the institution's name was changed to Museum Victoria. Throughout this book, the name Museum Victoria refers to the institution through all its various guises and names, except where a particular campus is referred to. It took almost a century for the now much expanded museum's natural history collections to finally move from their temporary home in the State Library. After numerous proposals and false starts, the new Melbourne Museum in the Carlton Gardens was opened in 2000.

Wolf Herring (*Chirocentrus dorab*), collected by François Laporte, Count Castelnau, French Consul in Melbourne (1862–1880).

Genetic and palaeontological research provides an unforeseen look at the evolutionary history of species, many of which are only known from museum collections. An eighteenth-century museum director might be horrified by the evolutionary bent of modern museum researchers, but would be mollified to see their commitment to research and education continue unwavering through the dedication and efforts of scientists working at the heart of the institution. It is to these scientists, their collections and their research, that this book is dedicated.

This book is a journey through the back rooms of a museum – through a natural history collection – inspired by a handful of the strange and curious creatures collected there. It is a journey through Australian natural history, where our past collides with our future. It travels into our forests, mountains, rivers, oceans and islands; into the depths of the earth and the farthest reaches of outer space. These stories take us into our historical past, our evolutionary past and our geological past and beyond, to the future. It is not a comprehensive overview of the biological sciences; rather – like a scientist's mind – it is a voyage both eclectic and curious, where disconnected and disparate objects unite to reveal unexpected patterns. Scientists do not think like textbooks – they focus at a seemingly impenetrable level of detail before pulling back to survey the broader intellectual landscape; they draw on a stored miscellany of facts and observations loosely attached to an array of possible interpretations. In chasing an inconsequential and seemingly unrelated topic, they often serendipitously stumble across the key that fits the missing piece of the puzzle that has occupied their research for years. What follows are the stories of eleven such 'keys' from deep in the collections of Museum Victoria.

2 A BEAST
NAMED SU

Bark painting of barramundi and kangaroo from Western Arnhem Land.
Indigenous Collections, Museum of Victoria.

IN THE SIXTEENTH century Europeans began a wave of sea explorations which would ultimately circle the globe, into an unknown world populated by mythical and frightening beasts. As the world slowly revealed its secrets, these creatures were replaced by even stranger and more peculiar animals, quite different from the familiar beasts of Europe. And nowhere were these creatures so anomalous as those found in Australia.

Kangaroos burst into the European imagination in the late 1700s after English accounts of the expedition of Captain James Cook (1728–1779) depicted the extraordinary animals in their pages. It was Cook's shipboard naturalist (and financier) Joseph Banks (1743–1820) who first wrote the name 'kanguru' in his diary of 14 July 1770, understanding this to be the Aboriginal name for the strange creature they had just shot. The capture of this creature ended weeks of speculation for the men of the *Endeavour*. On 22 June, Banks reported that the crew had seen 'an animal as large as a greyhound, of a mouse colour and very swift'. A few days later Cook saw a similar animal and concurred with the earlier description but noted that 'I could have taken it for a wild dog, but for its walking or running in which it jumped like a hare or deer.' Soon, everyone on the ship seems to have seen the animal except Banks, but eventually he too saw one, noting that he could not determine its mode of locomotion due to the long grass. No doubt this explains why the others failed to observe the obvious, if unexpected, feature of the kangaroo; that it is bipedal rather than moving on four legs like the greyhound or deer.

The first macropod shot by Cook's crew was one of the larger kangaroos (which can attain weights of over 80 kilograms and stand up to 2 metres tall). In the coming centuries, much attention was paid to this strange large marsupial. Every aspect of the kangaroo's physiology drew comment. The elderly writer Samuel Johnson (1709–1784) gravely mimicked its remarkable method of locomotion, to the astonishment of his Scottish hosts. The English essayist Charles Lamb (1775–1834), commenting on its shortened forelimbs, suggested that they were for picking pockets (perhaps a comment on Australia's convict settlements). As the naturalist George Bennett (1804–1893) was later to reply, the ability to pick pockets was entirely appropriate for a member of the only group of animals to have their own pockets to pick. The pouch is, of course, the defining feature of any marsupial, and the kangaroo's pouch attracted the most comment. Much was made of the kangaroo's devotion to its young, which retreat, even when quite large, to their mother's pouch whenever danger threatens. Even when a younger sibling prevents the young

A kangaroo illustration from Hawkesworth's 1773 edition of the *Endeavour* journal, based on a work by George Stubbs which was drawn from an inflated skin. This illustration provided the model for numerous other English illustrations.

kangaroo from climbing into the pouch, it continues to suckle on an enlarged teat and remains in its mother's company.

By the time the first live kangaroos arrived in London in 1791, the scene was set for enormous public interest in the strange pouched creatures from the other side of the globe. With their eyes on commercial success, both the English publishers and the exhibition hall promoters ensured that history would be littered with images of kangaroos from the late 1700s. To the casual viewer, it can often seem that kangaroos, like Australia itself, were discovered by the English. But like the Anglocentric view of the discovery of Australia (still promulgated in classrooms and popular media today), the history of European, and scientific, understanding of kangaroos and their marsupial relatives is much longer and far more complicated. Just as Cook's 'discovery' of Australia can be challenged by earlier Dutch and Portuguese explorers, so too can the notion that the English discovered the kangaroo. And the patterns of kangaroo 'discovery' follow remarkably similar lines to the patterns of Australia's discovery by Europeans.

WAS IT A WALLAROO?

It was the French biologist Georges Buffon (1707–1788) who initially classified the kangaroo *Jerboa gigantea*. His arch rival, the Swede Carolus Linnaeus (1707–1778), had other ideas, classifying it as *Didelphus gigantea*, in the same genus as the American opossums. The vast difference between the American marsupials and the Australian species, however, seemed to demand a separate name. George Shaw (1751–1813) from the British Museum proposed the genus *Macropus* in 1790, naming the species *Macropus giganteus* – the name still applied to the Eastern Grey Kangaroo. Shaw provided the first detailed description of the species (albeit on rather limited data) and thus lays claim, taxonomically, to being the first European to describe a kangaroo; indeed, the full taxonomic designation of the species reads *Macropus giganteus*, Shaw 1790.

But the specimens sent to England by Banks and Cook may not have been Eastern Grey Kangaroos at all. Little remains of Cook's three original specimens, which were not complete and may not all have been the same species. Cook's specimens were shot near what is now Cooktown on the north Queensland coast. Four other kangaroo species are found there, which could be confused with the Eastern Grey – the Antilopine Kangaroo (*Macropus antilopinus*; Plate 5), Wallaroo (*M. robustus*; Plate 2), Whiptail Wallaby (*Wallabia parryi*) or Agile Wallaby (*W. agilis*). A drawing of one of the skulls seems to be of a Wallaroo, but unfortunately the skull itself was destroyed when the Museum of the Royal College of Surgeons was bombed during World War II. Banks originally described these animals as 'kanguru' because that was the name applied by the Guugu Yimidhirr people of the Cooktown region. Banks' annotation may have been of the word 'ganurru', which is used to refer to large black or grey kangaroos, and specifically to male Wallaroos. (Europeans later used the word kangaroo in conversations with other Aboriginal people, leading the Baagandji people of Darling River region to believe that 'gaangurru' was the English word for horse.)

If Banks' kangaroos were actually Wallaroos then, under the strict conventions of scientific nomenclature, the Eastern Grey Kangaroo is not entitled to its scientific species name of *giganteus*. By rights, that name should be applied to the Wallaroo. But the name seems to have been retained because of its historical link to Cook.

When the first specimens of Banks' kangaroos arrived in Britain in 1771, they were far from being the first marsupials to be seen there. Indeed, Bank's detailed descriptions were not even the first written accounts of Australian marsupials or Australian macropods. Just who did discover, draw and describe the first kangaroo and introduce this very different creature to the European imagination?

The first known European drawing of a macropod dates from 1705, when the Dutch artist Cornelis de Bruijn (1652–1727) observed a captive wallaby (now known to be the Dusky Pademelon, *Thylogale brunii*) in General Joan van Hoorn's garden in Java.

> The Filander, which has hind limbs much longer than the fore, is nearly the size of, and possess nearly the same form as, a large rabbit. The head resembles that of a fox, and the tail is pointed; but the most extraordinary circumstance is that the female has a bag-like opening in the belly into which the young ones enter, even when they have attained a considerable size. They are often seen with head and neck thrust out of this bag.

Descriptions of Australian macropods, however, date back even further. In 1699, English privateer William Dampier (1651–1715) observed another west coast wallaby, which he described as being a type of jumping raccoon 'with short forelimbs'. It seems likely that this species was a Banded Hare Wallaby (*Lagostrophus fasciatus*), a pretty little creature with silver and chocolate brown bands across its rump. Just over a century later this same species was among the earliest macropods to be scientifically described and named, when specimens from Shark Bay in Western Australia were sent back to France by the naturalist François Péron (1775–1810).

The Quokka (*Setonix brachyurus*), a small wallaby only half a metre high, was described even earlier. Although it is not a member of the 'true' kangaroo genus Macropus, the Quokka does belong to the wider Macropodidae family, including tree-kangaroos, rock wallabies and others. Rottnest Island, where the majority of Quokkas are now found, was named by Willem de Vlamingh (1640–unknown) in 1696. His comrade Witson described 'rats nearly as big as cats' with 'a pouch below their throats into which one could put one's hand, without being able to understand to what end nature had created the animal like this'. His countryman Samuel Volkerson had 'thirty-eight years earlier' described the same species on this island less accurately as resembling 'a civet-cat, but with browner hair'. Although Vlamingh 'had great pleasure in admiring this island, and where it seems that nature has denied nothing to make it pleasurable beyond all islands I have ever seen,' further exploration of the Western Australian coast

The tiny Banded Hare Wallaby may be the last survivor of the massive Sthenurinae or short-faced kangaroos, which weighed upt to 200 kilograms and stood up to 3 metres high. These giants disappeared about 30,000 years ago; in the last 5000 years their diminutive descendant has also disappeared from across southern Australia and is now found only on two small islands in Shark Bay, WA.

C. A. Lesueur (1807).

seems to have left the Dutch expedition underwhelmed On 21 February 1690 another crew member, Mandrop Torst, reported that their ships fired their cannons 'as a signal of farewell to the miserable South Land'.

The first known European description of an Australian marsupial was made in 1629 by Francisco Pelsaert (1590–1630), captain of the doomed Dutch treasure ship the *Batavia*. Pelsaert's ship, laden with jewels, gold and silver, was bound for Java when a storm sent it south, wrecking it on an unknown coast. More than 130 crewmen and 30 women and children landed on a cluster of small islands near the reef where their wrecked ship lay. Pelsaert went in search of water, which was in short supply; finding little, he eventually headed north for Java, returning to rescue his crew and passengers. But in the intervening three months murder and mayhem had ensued among the survivors. Men, women and children had been brutally slain by other crew members. Rather than merely rescuing grateful survivors, Pelsaert was required to put on trial and execute the culprits while also retrieving his employer's valuable possessions.

Under these difficult circumstances, Pelsaert provides us with a detailed and accurate account of the Tammar Wallaby (*Macropus eugenii*):

on these islands there are large numbers of Cats, which are creatures of miraculous form, as big as a hare; the Head is similar to [that] of a Civet-cat, the fore-paws are very short, about a finger long. Whereon there are five small Nails or fingers, as an ape's fore-paw, and the 2 hind legs are at least half an ell long [34 cm], they run on the flat of the joint of the leg, so that they are not quick in running. The tail is very long, the same as a Meerkat [or a lemur]; if they are going to eat they sit on their hind legs and take the food with the fore-paws and eat exactly the same as the Squirrels or apes do. Their generation, or procreation is Very Miraculous, Yea, worth to note; under the belly the females have a pouch into which one can put a hand, and in that she has her nipples, where have discovered that in there their Young Grow with the nipple in mouth, and have found lying in it [the pouch] some Which were only as large as a bean, but found the limbs of the small beast to be entirely in proportion, so that it is certain that they grow there at the nipple of the mammal and draw food out of it until they are big and can run. Even though when they are very big they still creep into the pouch when chased and the mother runs off with them.

THE NIPPLE ORIGIN HYPOTHESIS

Pelsaert's suggestion that young macropods somehow grow out of their mother's nipples has been a pervasive myth. The question of marsupial generation occupied the letters column of the Melbourne *Age* for several weeks in 1870, with many correspondents supporting the 'nipple-origin' hypothesis. This strange idea has been dispelled fully from the public imagination only in recent years by television documentaries showing the birth and early journey of infants into their mother's pouch.

We now know that marsupials give birth to tiny, underdeveloped young that 'swim' through their mother's fur up to the pouch, which contains her teats. The neonates attach to a teat, which then swells inside their mouth, preventing their accidental dislodgement. The young remain attached until they reach the size at which (in most other species) they would be born. When kangaroos were first bred in England, 'birth' was measured from the time the joey's head first emerged from the pouch, and the emergence of the young joey from the pouch is still termed its second 'birth'.

Gradually, as they attain independence, the young marsupials leave the pouch. The offspring of many tree-dwelling marsupials leave the pouch quite young and climb onto the mother's back (like the opossum) or remain in a den (like the carnivorous quolls) while she searches for food. In the larger ground-dwelling macropods, the pouch young do not impede movement and may remain in the pouch longer, even returning to it when they are quite large.

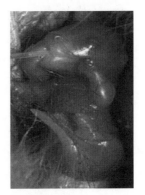

Kangaroo neonate.
M. Renfree.

Although marsupials are synonymous with Australia and indeed, the Australian region is the great centre for diversity of marsupials, they are not unique to this country. Marsupials are also found in South America, with the Common Opossum (*Didelphis virginiana*) extending into North America. The King of Spain was presented with the skin of a Southern Opossum (*Didelphis marsupialis*) as a curio from Brazil in 1500 by a returning explorer, Vincente Yáñez Pinzò. This opossum was a female with young, and her pouch drew a great deal of attention. This strange new creature was depicted in the lower left-hand corner of Waldseemüller's world map of 1516. The distinctly dog-like image (probably drawn from accounts rather than from life or a specimen) seems to have been the model for subsequent images during the sixteenth and seventeenth centuries. Konrad Gesner's massive *Historie Animalium* in 1558 described the opossum as a 'monkey-fox' or simivulpa on account of its fox-like face and monkey-like hands and tail. In the 1600s the Reverend Edward Topsell's *Historie of Foure-footed Beastes* referred to the strange creature as a 'Wilde Beast named Su' from Patagonia. Over time, the misshapen creature acquired features more in keeping with an actual opossum. Topsell's image has shorter legs than Waldseemüller's and carries its young on its back. In subsequent editions of Topsell, the mother's squirrel-like tail was refined to a prehensile form to which her offspring cling with their own tails. Her puggish face was also replaced with a more accurate opossum-like head.

The marsupial pouch and other strange features of its reproductive system attracted a great deal of attention in Europe and America. Some suggested that the opossum mated and gave birth through its nose, an idea arising from the observation of the male opossum's bifurcate (or two-pronged) penis and the observation that the female opossum licks its pouch and vagina while giving birth. The bifurcate penis and dual uteri of the opossum also gave rise to the suggestion that the pouch was actually an external womb (with another internal womb) but Edward Tyson's thorough anatomical work in 1690 laid these myths to rest. Nonetheless, in 1735 Carolus Linnaeus christened the opossums *Didelphis* or 'two-wombed'. Linnaeus also seems to have been unaware of an increasing number of accounts of marsupials from lands other than South America.

So even in the early eighteenth century macropods, and certainly marsupials, were known to Europeans, although they remained a novelty. Not only were the Dutch among the earliest European visitors to Australia but, not surprisingly, they were among the earliest to describe its unique fauna. But just as the Dutch cannot rest easily on their laurels as

The Reverend Edward Topsell's 'wilde beast named Su', or opossum (1607), in a gradual process of transformation from a dog-like image to the more familiar marsupial form we know today.

PLATE 1. Bark painting of barramundi and kangaroo from Western Arnhem Land,
acquired for the Indigenous collections of Museum Victoria before 1913.

PLATE 2. Eastern Wallaroo (*Macropus robustus*, formerly *Osphranter robustus*).
John Gould, *Mammals of Australia* (1863), vol. II, Plate 11.

PLATE 3. Grizzled Tree Kangaroo (*Dendrolagus inustus*).
John Gould, *Mammals of Australia* (1863) , vol. II, Plate 50.

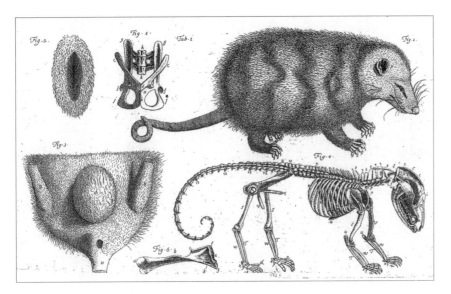

The anatomy of a
dissected opossum.
E. Tyson, (1698).

the first European discoverers of Australia, neither can they be assured of
the claim to have 'discovered' the macropods. The contender for that
crown, as with the discovery of Australia, is Portugal.

Evidence of Portuguese activity in Australia is provided not by the
Portuguese themselves, but by the French. In the early 1500s the French
city of Dieppe was a major centre for European cartography, and a series
of world maps was produced there between 1536 and 1566, known
collectively as the Dieppe maps. The first of these ornate and richly
illustrated maps was created by Pierre Descalier for the King of France to
give to his son, the Dauphin. The Dauphin map, like its fellow Dieppe
maps, contains incontrovertible evidence of information which could only
have come through Portuguese explorations – Portuguese place names as
well as details of regions which at the time were vigorously protected
from foreigners by the Portuguese. For this reason, the Dieppe maps tend
to be referred to as Portuguese rather than French. How the French came
to have this fiercely guarded information remains contentious, but it seems
generally agreed that much of the maps' content, particularly in relation to
the Australasian region, is Portuguese in origin.

The last Dieppe map was produced by Nicolas Desliens. Unlike his
contemporaries' maps, the Desliens map is sparsely decorated. Desliens
seemed little interested in the mythical or imaginary, and it seems
reasonable to assume that his representations are as accurate a summary of
the state of European knowledge at the time as it was possible to make.
The Desliens map is notable for its relatively accurate portrayal of some
aspects of the Australian coast, a land which Desliens labelled with a
Portuguese flag.

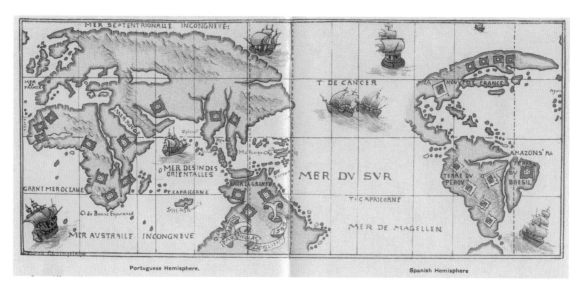

Nicolas Desliens' 1566 map of the world divided into Spanish and Portuguese territories and featuring the coast of 'Java La Grande'.

At the time of Desliens' death in the last half of the sixteenth century, atlases had become popular. In many cases, these replicated the Dieppe maps but incorporated more and more errors. But the atlas produced by Gerard and Cornelis de Jode was an exception and reveals the most convincing depiction of Australia of any of the maps of the time. This atlas, the *Speculum Orbis Terrae*, was published twice, once in 1578 and again in 1593.

On the frontispiece of de Jode's atlas was a cartouche with an animal on each of the four corners. A horse, a camel and a lion are unmistakable. But on the fourth corner, at the bottom right-hand side, is an unknown creature. It looks a little bit like a camel, with a somewhat reptilian head on a long neck. It has no tail, but its hind legs, curved beneath it as it lies on its scroll, are powerful and between its forepaws, attached to its chest, is an unmistakable pouch containing two young.

Kenneth McIntyre, who has collated much of the controversial evidence for a Portuguese presence in Australia in the sixteenth century, argues that this mystery animal is a kangaroo and it represents Australia. The other three animals each represent one of the known continents – the horse for Europe, the camel for Asia and the lion for Africa. Could this strange distorted creature be the first use of a kangaroo as a symbol for Australia?

The image may or may not be a kangaroo, but it is unmistakably some kind of marsupial. Does the depiction of a marsupial on a sixteenth-century map suggest that the Portuguese had a greater knowledge of Australia than they were letting on? De Jode's atlas also depicted a southern continent separated by a narrow strait from New Guinea. On this land he

OPPOSITE: The frontispiece of *Speculum Orbis Terrae* (1593) by the Flemish mapmaker Gerard de Jode.

SPE-
CVLVM
ORBIS
TERRÆ

ANTVERPIÆ.
Sumptibus Viduæ et Heredū Gerardi de Iudæis.

drew a lone archer preparing to shoot a winged dragon which stands upright on its back legs, balanced by a long tail and small forelimbs. Many of the animals drawn on these atlases were entirely mythical and imaginary. It could be argued that the pouched creature on de Jode's frontispiece is equally fanciful. After all, its pouch almost appears to be suspended around its neck, not on its abdomen as is the case with real marsupials. But even first-hand descriptions of marsupials were misleading. De Vlamingh described the Quokka as having 'a kind of bag or purse hanging from the throat upon the breast downwards'. Europeans had a great deal of difficulty finding the words to describe these bizarre new creatures and fell back upon analogies with the familiar – cats, raccoons, rats, monkeys and foxes. Far from being limitless, our imaginations really do seem constrained by our experience. Perhaps Banks had the right idea when he refrained from drawing a parallel between a kangaroo and any other animal on the grounds that 'nothing certainly that I have seen at all resembles him' (25 June 1770). De Jode was probably drawing from a written or spoken account (second- or third-hand) of an animal, rather than copying an actual creature from life, or even from another drawing. In truth, de Jode's 'kangaroo' looks more like a kangaroo than any of the early opossum images resemble opossums – but that is no great claim.

Given that the Europeans knew about marsupials from the Americas, it is possible that de Jode meant the unknown creature to represent a South American marsupial (albeit considerably larger than any opossum). However, the axes on which these animals are positioned are generally interpreted as being indicative of direction. In de Jode's maps, South America is to the west of Africa, not to the east as is this marsupial. Australia seems to be the logical location to be represented by this creature.

Australia is not the only country in the bottom right-hand corner of de Jode's world map to have marsupials. The natural range of the marsupials extends up into New Guinea and its neighbouring islands. Human activity has probably spread their distribution even further west (confounding efforts to determine lines of biogeographic demarcation between Asia and Australia). But at least in recent human history marsupials have been found as far north (and west) as the Spice Islands of the Moluccas, a hub of sixteenth-century Portuguese activity.

The Portuguese established themselves in the islands north of Australia in the early 1500s, taking over the maritime trade routes in South-East Asia, long controlled by the Chinese and Arabs. Between the eleventh and thirteenth centuries, Arab traders apparently brought back stories of long-tailed animals which suckled their young in a pouch, but it is unclear whether the Portuguese took on such fragments of information.

THE PORTUGUESE IN AUSTRALIA

Portugal's strength as a world power in the 1500s stemmed largely from its maritime prowess, developed under the far-sighted and enlightened guidance of Prince Henry the Navigator. New and old methods of bluewater navigation were explored, synthesised and standardised. Prince Henry's revolution was one of knowledge and he gathered about him men who had that knowledge – of ships, of the sea, of mathematics and science, of foreign lands and geography and, most important of all, of maps. Using a specially developed ocean sailing vessel, the caravel, Prince Henry's vision swept aside the supposed boundaries of the oceans, with their boiling seas, swirling serpents and certain death, and set course to take Europe to the rest of the world.

The history of Portuguese activity in the Australasian region is shrouded in intrigue, espionage and suspicion. In a climate of exploration and growing maritime trade, the discovery of the Americas and the possibility of a route to the East Indies sparked intense rivalry between the reigning maritime powers, the Portuguese and the Spanish. Neither country was prepared to be entirely frank about what it did and did not know about the new worlds. But borders had to be drawn and the Treaty of Tordesillas in 1494 divided the 'non-Christian' world down the mid-Atlantic, offering right of first refusal of any uncolonised lands to the Spanish west of Europe (the Americas) and to the Portuguese in the east (effectively Asia).

Although both parties seem to have understood that the Atlantic line of demarcation must have an obvious extension on the opposite side of the globe, neither country was technologically, nor politically, able to offer a clear Pacific line of demarcation. Inevitably, this (and other ambiguities) led to disputes, with the Spanish seeking to approach the valued East Indies via America, through 'Spanish' waters. As the weaker power, Portugal relied heavily on papal edict and international treaty to defend its acquisitions. And not surprisingly, secrecy was a major component of its strategy. Portugal was not about to lay its discoveries of new lands before the covetous eyes of its opponents, particularly when Portugal had little prospect of defending them in a show of force against the powerful Spanish.

For most of the sixteenth century the Portuguese dominated the spice trade from the Moluccas, and retained a colony in the area (East Timor) until 1972. In 1526, the Portuguese sailor Jorge de Meneses was wrecked on one of the northern New Guinean islands (if not New Guinea itself). In 1525, another sailor, Gomes de Sequeira, discovered the southerly island of Tanimbar. The nearby Kai and Aru Islands also have Portuguese connections. Until 1529, the Portuguese kept their knowledge of New Guinea (and perhaps Australia) secret. These lands lay in Spanish territory and Portugal was anxious not to offer any excuse (such as accidental infringement of Spanish territory) for the Spanish to retaliate by encroaching into Portuguese territory. The Spanish would not have needed much provocation to justify seizing the valuable Spice Islands, which lay just west of the line of demarcation. As a consequence, Portugal kept silent on the accidental incursions of their storm-tossed mariners to New Guinea and Tanimbar. In 1529, however, Portugal made a deal with the cash-strapped Spanish and the line of demarcation was pushed east to the edge of Australia. For the next fifty years (until the demise of the Portuguese empire) Portugal had free run of New Guinea and its islands and, in fact, Australia. The extent to which they made use of this new-found freedom is, however, less clear. It seems more likely that the Portuguese merely used this extra space as a buffer against Spanish incursions rather than a resource to exploit. Nonetheless, there was nothing to prevent the Portuguese from entering these waters, accidentally or deliberately.

The potential for the Portuguese to meet a macropod was surely high. Marsupials, in the form of the cuscus (*Phalanger orientalis*) were certainly known to the Portuguese by the mid-1500s. Long-unpublished manuscripts from the Jesuit Library in Seville record a description of a cuscus from the Station Captain on Ternate in 1536–1540 (recently published by Father Hubert Jacobs):

> Some animals resemble ferrets, only a little bigger. They are called *kusus*. They have a long tail with which they hang from the trees in which they live continuously, winding it once or twice around a branch. On their belly they have a pocket like an intermediate balcony; as soon as they give birth to a young one they grow it inside there at a nipple until it does not need nursing any more. As soon as she has borne and nourished it, the mother becomes pregnant again. The people eat them like rabbits, seasoned with spices.

Despite the secrecy surrounding this material, reports of Moluccan animals like opossums began to appear in the literature by the seventeenth century. In 1606, an account of the journey of the Spaniard Luis Vaez de Torres through the strait between Australia and New Guinea which now bears his name included a description of a small dog-like animal with a long scaly tail – probably a cuscus. Could the Portuguese also have encountered macropods in their distant colony?

The western half of New Guinea and the Moluccan islands now lie in the territory of Indonesia. Until the last twenty years, our knowledge of the mammal fauna of this region was patchy, often consisting only of the isolated records from earlier centuries. Tim Flannery has done much to rectify this deficit by collecting, collating and correcting records of mammals in the region and publishing them in a series of books. According to this work, Indonesia can claim three species of macropods as well as numerous other marsupials, particularly species of cuscus, which have probably been spread through the island chain by humans. One of the local macropods, the Grizzled Tree Kangaroo (*Dendrolagus inustus*) is commonly kept as a pet by locals (Plate 3). Although its natural distribution can be confirmed only for northern New Guinea and the New Guinean island of Japen (on its northern coast), others claim to have seen it on the islands of Waigeo, Salawati, Great Kai Island and the Aru Islands.

In modern times, Grizzled Tree Kangaroos have been found in human company on islands as far west as Hamelhera, an island which was once the heart of Portugal's power base in the Moluccas. There is no reason to suspect that the transportation of macropods between the islands is a recent phenomenon. De Bruijn's wallaby in Java (where no marsupials are found naturally) was a Dusky Pademelon (*Thylogale brunii*) or Aru Island Wallaby from a captive population bred for food. De Bruijn reported that the tables fairly groaned under the weight of 'Aru rabbits' on special occasions. The Dusky Pademelon also occurs naturally in southern New Guinea and the Kai Islands.

The third Indonesian macropod is the Brown Dorcopsis (*Dorcopsis muelleri*). This species is also found on the islands surrounding New Guinea. But these island populations were clearly not transported by humans. The Dorcopsis of Misool, Japen and Salawati are quite distinctive in appearance, suggesting that they have been isolated from one another for at least 10,000 years and probably since sea levels rose and cut off the islands from mainland New Guinea.

These wallabies are considerably larger than the marsupials and macropods described by Europeans prior to Cook. While not attaining the stature of the large plain-dwelling Australian kangaroos, they all stand well over half a metre high and have the characteristic macropod shape and appearance. The simplest explanation for de Jode's marsupial is that it was a kangaroo of some kind, but not that it is the first drawing of an Australian kangaroo. I think it is much more likely to reflect reports (if not accurate images) of Grizzled Tree Kangaroos, Dusky Pademelons or a Brown Dorcopsis. The first kangaroo to arrive in the imagination of Europe was not an Australian at all, but New Guinean.

Of course, from an indigenous perspective all of these arguments about historical precedence seem like fleas arguing about ownership of the dog's back. Without question, Aboriginal people were the first to draw, describe and understand kangaroos (Plate 1). A silhouette of a kangaroo in the lower Darling River region of New South Wales was engraved in the rock between 6000 and 10,000 years ago. Aboriginal Australians have lived in this country for at least 40,000 years, perhaps back to the dawn of humanity itself. Their knowledge of kangaroos is extensive and part of a complex relationship between the land, the kangaroos and the people. Science, with its European origins, sometimes seems as if it is only beginning to come to terms with indigenous sources of knowledge, but in fact, as the next chapter explores, the relationship between professional scientists and indigenous knowledge has a much longer and surprisingly productive history.

3 LOCAL
KNOWLEDGE

A Lesser Bilby (*Macrotis leucura*), collected by W. B. Spencer with the assistance
of people from the Mutitjulu Community of central Australia.
Mammalology Collection, Museum Victoria.

The history of science is often presented as a history of great men. Aristotle, Galileo, Newton, Darwin, Einstein and others all represent epochs in the development of science as means of acquiring knowledge about the world, and it is tempting to summarise the history of science through a series of individual biographies.

But science is the accumulation of knowledge collected by thousands of people in different places and at different times. Despite the dominance of a handful of 'great men', their work rests on the back of a myriad of other lesser mortals — not just on the shoulders of giants, as Newton put it, but also unnamed collectors, unpaid wives and unacknowledged 'amateurs'. In Australia in particular, where there is a tendency to denigrate even the work of the 'great men' of Australian science, the achievements of these unofficial contributors to science rarely rate a mention. Professional scientists leave a written paper trail of their work and contributions, preserved for all posterity in the scientific literature, but tracing the contribution of non-professional scientists can be like trying to track footprints in the sand on a long deserted beach.

Where the contribution of amateur scientists is understated, the contribution of indigenous knowledge to the development of the biological sciences is almost completely hidden. And yet the expertise of Aboriginal people in the 'discovery' of Australia and its wildlife to European science was absolutely critical. Early explorers often would not have even survived without the generous assistance of local people. George Bass (1771–1803) regarded his ability to communicate in seven indigenous languages as vital to his success as an explorer. Most explorers recruited the assistance of indigenous people to guide them through new lands, sometimes by mutual agreement, often by force. Explorers' diaries record how they captured locals and forced them to show the way to water and find food. Such enforced recruits often escaped in the night, and subsequent explorers were surprised by the lack of enthusiasm their arrival generated.

Many early collecting expeditions were not so much for the direct collection of specimens by the expedition leaders themselves, but trading arrangements with indigenous communities. William Blandowski (1822–1878), for example, acknowledged the assistance of his 'friends the Yarree Yarree Aborigines' in providing all of the twenty-six mammal species collected during his expedition along the Murray River for the founding of Museum Victoria in 1857. In return Blandowski provided flour, sugar, tea, blankets and clothing to the value of £220. His collecting methodology was later criticised by his assistant, Gerard Krefft, as relying too heavily upon the exertions of others. But many other collections of the time relied just as heavily upon indigenous expertise, though few were as generous in their acknowledgements as Blandowski.

WOMEN IN SCIENCE

Many women have played a significant, but often under-acknowledged, role in the development of Australian natural history. Many of John Gould's famous paintings of Australian birds and mammals were done by his talented wife Elizabeth (1804–1841; Plate 4), who travelled with him to work and care for their infant son. The efforts of Amalie Dietrich (1821–1891), who spent ten years collecting specimens in Australia for a German museum in order to support her daughter, are known more for their emotional effect on the literature of her daughter than they are for their scientific significance. Few students of pseudocopulation in wasps and orchids are aware of the prolific amateur naturalist Edith Coleman (1875–1951), who discovered and described the phenomenon in the Victorian grasslands near her home. Jeanne Baret, the Frenchwoman who travelled as valet to the naturalist Philippe Commerson (1728–1773) on Bougainville's Pacific expedition in 1786, barely figures in the literature except as a romantic echo in genre novels. Sir John Franklin features in many a dictionary of biography as Arctic explorer and lieutenant-governor of Van Diemen's Land, but few document his wife Jane (1792–1875), who worked tirelessly to support scientific studies of native flora and to promote humane treatment of convicts and indigenous Australians and founded Tasmania's Natural History Society. Like George Eliot's Dorothea, these women spent their strengths 'in channels which had no great name on earth … [and] rest in unvisited tombs', but the world is an incalculably better place for their 'unhistoric acts'.

Elizabeth Gould, wife and illustrator to John Gould.

William Blandowski
receiving collections from
members of the local
Aboriginal community at
his campsite on the
Murray River.
Haddon Library, Cambridge
University.

With the disruption of Aboriginal culture through European colonisation, much of that knowledge has been lost, particularly on the east coast of Australia. In many cases, the only record of indigenous knowledge of the pre-European environment is in word lists preserved by early collectors. The local police superintendent and magistrate in Gippsland, Alfred Howitt (1830–1908), recorded the names and collected many specimens of species no longer found in the area. Even by the early nineteenth century it was apparent that Europeans had done a great deal of damage to Victorian Aboriginal societies and that much information about native species was being lost. Efforts to compile indigenous names for species was part of an erratic attempt to record such cultural information before it was lost entirely. Ironically, that 'cultural' information now forms the earliest known faunal lists for areas of Victoria. For example, Howitt recorded the Braiakaulung word for the White-footed Tree-Rat (*Conilurus albipes*; Plate 6). This species is now presumed extinct; fossil evidence suggests that it was common across the south-east but began to decline around the time of European colonisation. The fossil and linguistic evidence provides important information on the original distribution of an extinct species.

NAME WHAT YOU EAT

Classification is a pragmatic task, often applied to large animals and food. While Arfak highlanders in New Guinea readily identify different bird species (an important source of food and decoration) they do not distinguish between different ant species which are not utilised. Central Australian Aboriginal people like the Pitjantjatjara, however, have a far more detailed discrimination of invertebrates, particularly food species like honey ants, bees and witjuti (witchetty) and bardi grubs. Similarly English-speakers have a great diversity of words for domestic species like cows (distinguishing between males and females, young and old, castrated and intact) but happily categorise over 3000 different species of Australian Formicidae as 'ants'.

Function and behaviour are common criteria in many Aboriginal taxonomies. The people of Groote Eylandt in northern Australia seem to divide all living animals into three groups – those that live in the sea, those that live on land, and winged creatures. While further divisions take these classifications down to specific levels, the most detailed descriptions do not always correlate with scientific species names (Plate 5).

Western taxonomic systems use a hierarchy of names (both scientific and common) loosely based on an animal's appearance, rather than its function or behaviour. For example, a Southern Emu-wren is one of two types of 'emu-wren', which are generally grouped together with all other small, long-tailed birds or 'wrens', which in turn are part of a larger group known as 'birds'. Scientifically, the Southern Emu-wren has a designated species name of *malachurus*, preceded by the name of its genus *Stipilurus* (which it shares with the other emu-wren). Both species belong to the Australian warbler family (Maluridae), which belong to the order Passerine (perching birds), which in turn is a subset of the class Aves which includes all birds.

Many scientists and collectors recorded names for categories of animals that had no sensible counterpart in the local language. The surviving records of Victorian Aboriginal languages suggest that there was no general word for 'lizard' or 'snake' as collectors were often given the name of the most common snake species instead. Indigenous knowledge is also individually acquired and some members of a community know more than others – what you are told will depend on whom you ask and who you are, just as it does in any other culture.

A Braiakaulung name for the Tasmanian Pademelon (*Thylogale billardieri*) confirms that this species, now restricted to Tasmania, was once abundant on the mainland as well. Similarly, local names for the Pied Goose (*Anseranas semipalmata*), Brolga (*Grus rubicundus*) and the Australian Bustard (*Ardeotis australis*) all suggest much larger distributions for birds that are now found in abundance only further north.

The survival of indigenous languages and knowledge of native wildlife have been greater in other areas of Australia, allowing a more accurate map to be drawn of when animals became extinct. A dictionary of the Barngarla language from Eyre Peninsula in South Australia, compiled in 1844, suggests that Crescent Nailtail Wallabies (*Onychogalea lunata*), Long-haired Rats (*Rattus villosissimus*), Ringtail Possums (*Pseudocheirus peregrinus*), Burrowing Bettongs (*Bettongia lesueur*) and Lesser Bilbies (*Macrotis leucura*) were once present in the area, although they disappeared before any scientific surveys took place.

Museum collections are often the only record of fauna present in an area in colonial times. Valuable as this information is, it rarely represents a systematic sample of mammal species, and many collections took place after significant ecological damage had occurred in the region. Such collections therefore provide only a partial view of the ecosystem that existed prior to the arrival of Europeans. For example, the Flinders Ranges region of South Australia is currently known to have just twenty-two resident mammal species. But museum and historical records suggested that a further five mammals were once found in the area, including the Eastern Quoll (*Dasyurus viverrinus*), Burrowing Bettong (*Bettongia lesueur*), Long-haired Rat (*Rattus villosissimus*), a Hare-wallaby (*Lagorchestes* spp.) and perhaps the Black-footed Rock Wallaby (*Petrogale lateralis*). Owl pellets, under the right conditions, can sometimes preserve the hair and bones of mammals eaten by the owls for hundreds of years. Research on owl pellets has confirmed the existence of the Burrowing Bettong and Long-haired Rat in the Flinders, and added a further twenty-one names to the list. But while owl pellets are excellent for identifying the presence of those species commonly eaten by owls, they are less useful for identifying large mammals, or mammals which are not active at night when owls typically hunt.

In the 1980s the Adnyamathanha people of the Flinders Ranges in South Australia investigated the past and present distribution of mammals in the area with the assistance of specimens and skins from the South Australian Museum. Elders were able to identify and name many of the museum skins as animals that had once been found in their area, including three species not previously known to have lived there. Evidence from the cultural heritage and language of the Adnyamathanha people suggests that Lesser Bilbies (*Macrotis leucurus*), Brush-tailed Bettongs (*Bettongia penicillata*), and the Tammar Wallaby (*Macropus eugenii*) were once found in the area.

Tammar Wallabies are now found only in south-western Australia and on offshore islands, but they were once common on the mainland. They are recorded in numerous Aboriginal languages, many of which distinguish between the sexes and age classes, suggesting that they were common and important food animals. Brush-tailed Bettongs were similarly once common across southern Australia, but the arrival of cats and foxes (and perhaps rabbits) contributed to their extinction everywhere except a few isolated pockets in south-western Australia. Despite the fact that the Brush-tailed Bettong may not have been present in the Flinders Ranges since the early nineteenth century, oral traditions among the Adnyamathanha people have preserved a considerable amount of knowledge about it, including the characteristic diggings and burrows it made. In contrast, there is little information in the scientific literature on burrowing behaviour by this species.

The most significant species for the Adnyamathanha people is, however, the bilby, which is one of their totemic ancestors. As a consequence, a great deal of knowledge is handed down about the bilby, despite the fact that few living people have ever seen one in the wild. The remains of Greater Bilbies (*Macrotis lagotis*) have been found in owl pellets in the Flinders Ranges. But the Adnyamathanha distinguish between two totemic bilbies, the yarlpu (or Greater Bilby) and the warda, which is probably the Lesser Bilby (*M. leucura*). No other evidence substantiates the existence of the Lesser Bilby in this area, but the depth of indigenous knowledge about this animal strongly suggests that there were two species present – if not the Lesser Bilby, then some similar unidentified bilby.

Bilbies seem to have survived in greater abundance in inland regions of Australia, where cats and rabbits have made less of an impression. However, by the mid-1800s, Lesser Bilbies were rare even in central Australia. During the Horn Expedition to central Australia in 1894, Walter Baldwin Spencer mentions the existence of a second kind of bilby, which might well be the poorly identified Lesser Bilby. Again, most of Spencer's information on this species comes from indigenous sources, albeit through the medium of a local settler.

Spencer quotes notes made by Mr Byrnes at Charlotte Waters on the habits of the Greater Bilby (Urgǎtta), Lesser Bilby (Urpila), the Western Barred Bandicoot (*Perameles bougainville*, Tubaija) and Desert Bandicoot (*Perameles eremiana*, Iwurra):

[LEFT] Greater Bilby
(*Macrotis lagotis*).
G. Kreft.

[RIGHT] Western Barred
Bandicoot (*Perameles
bougainville*).
G. Kreft.

While the Urgåtta occupies the inner extremity of his burrow, the Urpila during the cold weather lies within a foot or so of the entrance of his, and only uses the inner chamber during the summer. This peculiarity is taken advantage of by the natives who spring upon the surface of the ground behind the Urpila breaking it in and so cutting off his retreat to the inner chamber. He is thus compelled to rush out through the entrance where a native is waiting to give him his quietus. The Urgåtta cannot be captured in this way, and has to be dug right out. Both species are nocturnal. The Iwurra and Tubaiji (Choeropus) are identical in their habits and build similar nests of grass and twigs in shallow, oval hollows scooped in the ground. They are captured in the same way, viz, by placing one foot on the nest pinning the animal down, and then pulling it out with the hand.

Sadly both the Desert Bandicoot, introduced by Spencer to science for the first time, and the Lesser Bilby, described just twenty years earlier, are now presumed extinct. The last sightings of both species by Aboriginal people in the area were in the 1950s and 1960s.

The reports of the Horn Expedition only occasionally mention collecting by Aboriginal people. Spencer was interested in the Aboriginal people he met during this expedition, an interest to which he was to dedicate much of his later career. But Spencer was inclined to see the indigenous inhabitants of the region only as the subject of investigation, rather than as a source of expertise or knowledge. The fact that the expedition relied upon 'black boys' to guide them is only incidentally mentioned. In contrast, white settlers providing information on the wildlife of the region are often named, even when clearly relying on information derived from indigenous sources. For example, Spencer reports on a local account that the Gould's Mouse (*Mus gouldi*) often nest communally, with multiple adult males, females and suckling young in the one burrow, and

PLATE 4. Mrs Gould's Sunbirds (*Cinnyris gouldiae*, now known as *Aethopyga gouldiae*) were named after talented artist Elizabeth Gould, who provided many illustration for her husbands's books.
A Century of Birds from the Himalaya Mountains.

PLATE 5. The Ritarango of Arnhem Land distinguish ten types of Antilopine Wallaroo on the basis of age, sex and behaviour. John Gould gave them all a single name (*Osphranter antilopinus* – now known as *Macropus antilopinus*) when he first described the species in 1842.
John Gould, *Mammals of Australia* (1863), vol. II, Plate 8.

PLATE 6. Howitt provided Museum Victoria with two specimens of the White-footed Tree-Rat (*Conilurus albipes*), which are catalogued as coming from central Australia. Although Howitt did collect in central Australia, there is no other evidence of this species ever having occurred there and it is likely that he actually collected them from Gippsland.
John Gould (as *Hapalotis albipes*), *Mammals of Australia* (1863), vol. III, Plate 1.

PLATE 7. The highly variable *Ctenotus* skinks have always been recognised as distinctive types by local Indigenous people, but have only recently been classified as such scientifically.
P. Roberts.

that local Aborigines can tell from the smell of the grass whether the burrow is in use or not. It is likely that the original source of this information, as well as additional material on the distribution of native mice in the area, was indigenous, rather than observed first hand by a settler.

Indigenous knowledge also contributed directly to the pioneering scientific investigations of the Horn Expedition. Spencer describes the speed with which local women could dig up the nest of the swollen-bellied Honeypot Ants (*Camponotus inflatus*) as much as 2 metres down, conspicuous on the surface only by a small hole. Incidental mention is made of their Aboriginal guides catching many of the common, but previously undescribed, Sandy Inland Mice (*Pseudomys hermannsburgensis*).

Walter Baldwin Spencer (1860–1920), anthropologist and professor of biology at the University of Melbourne and Director of Museum Victoria.

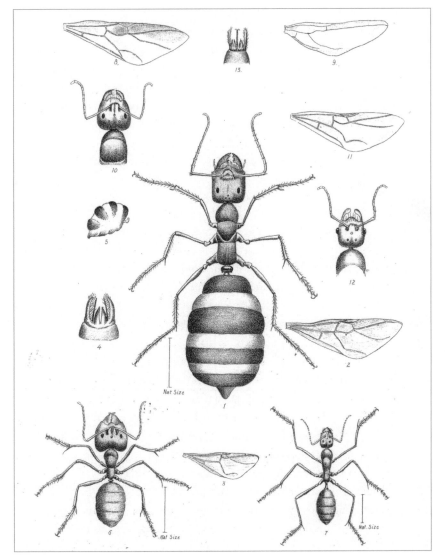

Honeypot Ants (*Camponotus inflatus*). W. B. Spencer (1896).

THE SPIDER AND THE
CHICKEN

Spencer had been puzzled by reports of a spider capable of 'barking' and was determined to solve the mystery. Barking spiders, which belong to the near legendary group of 'bird-eating' spiders, were the subject of numerous extravagant stories. In 1919, Mr Chisholm in North Queensland came across a chicken carcass which he claimed had been dragged 50 feet (16 metres). A barking spider was found attached to one of the chicken's legs, which it appeared to have dragged down into its burrow entrance. After such an impossible feat, the spider was probably fortunate Mr Chisholm relieved it of its adventurous meal before the inevitable indigestion.

The hotly debated 'barking spiders' (*Selenocosima crassipes*) were quickly found in burrows near the station by the expedition guides. After an evening listening at their burrows, Spencer concluded that the 'booming' call was probably produced by nearby nocturnal quail rather than the spiders themselves. Spencer did, however, concede that the spiders had effective stridulatory organs (like grasshoppers have on their legs) with which they could make a low whistling noise.

They also captured specimens of the rare marsupial hopping-mouse (*Antechinomys laniger*) or Kultarr and made the first identification of the Fat-tailed Pseudoantechinus (*Pseudoantechinus macdonnellensis*). Spencer was particularly intrigued by this small marsupial's ability to store fat in its tail: an adaptation, he noticed, which was also shared by fellow desert-dwellers like the Fat-tailed Dunnart, *Sminthopsis crassicaudata*.

In 1994, Museum Victoria, which holds some of the collections of the original Horn Expedition, hosted a return to some of the same collection areas. On this journey scientists were directly interested in obtaining local knowledge of species in the area, particularly the poorly understood invertebrate fauna. By collaborating with Pitjantjatjara and Yankunytjatjara people of the Muṯitjulu Community, the museum's scientists were able to contribute to a project designed to document indigenous knowledge of the invertebrates in the Uluru–Kata Tjuta National Park (Ayers Rock and Mount Olga). The indigenous community shared extensive knowledge of the behaviour and ecology of a wide range of invertebrate species, many of which are poorly understood by science. For example, they described the foraging behaviour of wolf spiders, centipedes and scorpions on the basis of

A SKINK'S TALE

In the case of many of the less well-known species (particularly reptiles, invertebrates and plants of less populated regions), indigenous knowledge still dramatically exceeds that of current scientific knowledge. For example, Western taxonomy once classified the highly variable *Ctenotus* skinks as representing a single species. Central Australian Aboriginal communities, however, have long recognised at least seven or eight different types, although many other names and classifications may have been lost or withheld. Differences were recognised on the basis of size, colour, habitat preferences and behaviour.

Science has now caught up with the older knowledge system, and more than a dozen different species are now recognised in the *Ctenotus* complex (Plate 7). Scientific knowledge of the skinks, however, remains restricted to basic morphological descriptions (the first step of understanding any living system). Aboriginal knowledge of the behaviour and ecology of these species still outstrips any strictly 'scientific' understanding of these species.

their tracks. They could identify the builders and structure of various invertebrate burrows, as well as the location of the resident within the burrow. They had extensive knowledge of the web-building behaviour of orb-weaving spiders. Much of this information is new to science and had not been documented.

Today, scientists are more aware of the direct value of indigenous knowledge, particularly for species that are little understood by science. Ken Simpson used Aboriginal knowledge to conduct broad-scale surveys of bilbies – documenting the sadly reduced range of the bilbies and when they were last seen in different areas. Other scientists have sought indigenous assistance in trapping local species. One research program was having no success in trapping an island population of Golden Bandicoots, until local people advised them to try 'sugarbag' or the honeycomb of the native bees. The success was immediate and the researchers were able to conduct scientific censusing work on the population.

Indigenous expertise also has great value in novel conservation research programs. Local women are assisting researchers in northern Australia to track the activities of feral cats in national parks. Using their valuable tracking expertise, local women can not only trace a cat's activity and identify what it did and ate during the night, but also lead researchers to the cat itself. The value of such tracking skills is also increasingly being recognised in African conservation programs. Scientists often use radio-tracking to follow the activities of far-roaming and nocturnal animals like leopards. This provides valuable information on ranging behaviour, but little else. Local San bushmen, however, are able to track the animals on foot the following day and describe what exactly they did (sleeping, interactions with other animals, hunting, species eaten). Adaptation to global positioning systems may allow the San to provide all the behavioural and location data in a form readily digestible by computers and scientists alike.

Australian Aborigines have certainly made, and continue to make, an enormous contribution to the scientific understanding of Australian wildlife, but this is not to suggest that indigenous knowledge itself is scientific. Science is a particular means of acquiring knowledge about the world which has proved extremely successful in a rapidly changing technological environment. Aboriginal knowledge is a successful means of acquiring knowledge about the world in a different cultural and physical environment. Science is a different knowledge system to that used by traditional Aboriginal communities, but that is not to say the two cannot interact and benefit from one another. Although both knowledge systems can beneficially filter into each other, they are not interchangeable; to suggest that they are in some way the same thing is to denigrate the value of both. 'Indigenous science' is not a term that has much value for anyone.

Notwithstanding conflicts of interest in the past between those working in the pursuit of science and the respect and well-being of Aboriginal communities, science and indigenous knowledge systems can and have interacted beneficially ever since they first came into contact. The immeasurable contributions of indigenous knowledge to biological science are undoubtedly matched by the benefits that science has to offer indigenous communities. Although poor health and social services continue to be a feature of many Aboriginal communities, this is a political failure to deliver scientific benefits, not a failing of science itself. Indeed, few would dispute or refuse the scientific benefits of medical research, communication technology or even the frequently denigrated Western agriculture. The history of science and Australian Aborigines may be chequered, but scientists and Aborigines, so long paraded as enemies, also have a harmonious and mutually beneficial history that is worthy of recognition.

Part 2

INTO THE FORESTS:
THE LAST 250000 YEARS

Thousand years ago

250 — Human brain reaches its maximum size

Australia becomes hotter and drier and sclerophyllous vegetation evolves

First modern humans evolve — 200

Charcoal remains suggesting possible human habitation laid down on what is now the Great Barrier Reef — 150

Evolution of the modern dog

Fires become more common in Australia — 100

Possible signs of human habitation in Australia

50 — 'King's Holly' begins to grow in Tasmania

Sea-levels rise, flooding Port Phillip Bay

Martian meteorite hits Antarctica

Australia separates from New Guinea — NOW

4 WATER, WATER EVERYWHERE

The Riffle Beetle or Water Penny (*Sclerocyphon* spp.)
is common to many freshwater river systems in Victoria
Terrestrial Invertebrate Collection, Museum Victoria.

MELBOURNE, AUSTRALIA'S SECOND-LARGEST CITY, suffers from little-sister syndrome in relation to its older, more glamorous relative, Sydney. Where Sydney is renowned for sunshine and crystal blue skies, Melbourne has an unenviable reputation for grey skies and rain. Sydney is built around one of the most spectacular natural harbours in the world, fringed with white surf beaches, while Melbourne is set inland on the edge of a wide, enclosed bay and murky river, both of which have the reputation for being highly polluted.

As is often the case, these reputations are undeserved and sometimes untruthful. Although Sydney's physical splendour can hardly be denied, like any large city it suffers terribly from pollution, particularly of its water. The famous Sydney beaches may glisten white in the sun, but sewage outfalls just offshore deny them the title pristine. They are often closed to swimming when bacteria levels reach dangerous levels. In 1998 Sydney's water supply was contaminated by high levels of cryptosporidium and giardia, and the entire city was ordered to boil drinking water to avoid contamination. The alert may have been somewhat premature and due to the introduction of more sophisticated testing techniques rather than any real bacterial increase. Nonetheless, there is no doubt that Sydney's water catchment, like that of many large urban centres, is under increasing pressure from urban development as well as agricultural runoff.

New York recently considered spending $8 billion on a filtration plant with an annual $300 million operating bill to meet its water demands. But there was an alternative. A mere $2 billion was found to be adequate to reclaim and restore the natural forested water catchment of New York's Catskill Mountains. Evidence of the success of natural and protected water catchment systems as a means of delivering safe clean water can be found much closer to home. Melbourne's water is delivered by one of the best natural systems of water collection and treatment in the world. It provides an example of the value of natural systems when planning makes use of their capacities, rather than abusing them.

Fresh water is arguably Australia's most critical resource, and water degradation is its greatest environmental threat. Irrigation schemes, artesian bores, deforestation of catchment areas and water pollution have all had catastrophic consequences for the Australian landscape. Lands have become unusable as salty groundwater has risen to the surface. Agricultural run off has resulted in toxic blooms that contaminate water supplies. Flood control has destroyed the revitalising and cleansing processes of the natural river systems, reducing many species to near extinction. Sydney's water problems are not unique, but symptomatic of the problems besetting the nation. The direct economic benefits of managing natural resources wisely are enormous, while the aesthetic and moral advantages are incalculable. Melbourne was initially sited on the banks of the Yarra River because it provided reliable fresh water. The wide mouth of the Yarra feeds into the salt water of Port Phillip Bay, but encroachment of salt was largely prevented by falls close to the river mouth. These falls, however, had a disadvantage: they prevented the movement of ships (the major means of transport in early Melbourne) and concentrated early residential development near the mouth, rather than along the river. Concentrated use in the one area increased pollution, and within ten years the waters of the Yarra were a source of constant complaint. The falls, in any case, were not as successful at preventing salt contamination as first thought; in dry years, Melbourne residents soon found themselves drinking brackish water, which was blamed for a high level of dysentery, diarrhoea and infant mortality. By the late 1840s industries such as tanneries and slaughterhouses added to the local flavour, discharging their wastes into the Yarra upstream from the water pumps which supplied the city's drinking water.

The development of regulated water supplies and sewerage coincides with the dramatic expansion of cities the world over in the late 1800s. London's Board of Works was formed only in 1855 to respond to the need for underground sewers, but it was not until the 'Great Stink' of 1858 that the upper reaches of the Thames were finally relieved of their burden of human excrement (which was diverted downstream). Similarly, Melbourne responded to the need for water supplies and sewerage in the early 1850s with the formation of the Sewers and Water Supply Board. The most enduring legacy of this body was the formation of the Yan Yean water supply, which rapidly repaid the debt of £800,000 incurred to build it. The Yan Yean reservoir was the brainchild of James Blackburn (1803–1854), a former London sanitary inspector transported to Australia for life for forging a cheque. Blackburn's skills as an engineer soon earnt him a pardon, and a job with the Melbourne City Council as City Surveyor. He selected the site for his alternative water supply on the basis of water purity, constancy, proximity to Melbourne and elevation (to allow gravitational flow). Ironically, given Blackburn's insistence on clean water as a matter of public health, he died of typhoid before the reservoir was completed.

Yan Yean reservoir was built on one of the tributaries of the Yarra – the Plenty River. A 9-metre embankment collected 29 million cubic metres of water, making it the largest artificial reservoir in the world at the time. Initially, the water quality was less than perfect. Lead jointed pipes resulted in several cases of poisoning and had to be replaced. Some of the Yan Yean

One of the many sources
of water contamination –
washing wool on the Yarra
Yarra, Melbourne, about
1885.

catchment drained from swamps carrying a heavy load of organic matter. Flow from the swamps was diverted. New aqueducts, weirs and secondary reservoirs redirected clean mountain water from the north flowing Silver Creek and Wallaby Creek south to Yan Yean, and Melbourne's water gradually improved. But the Yan Yean reservoir was surrounded by farmland and logged forests, which were soon recognised as a source of contamination. Remaining unalienated land surrounding the reservoir was reserved for water catchment, and over the next fifty years all habitation, farming and logging ceased in the catchment region. Fear of typhoid, which was known to be water borne, plagued Melbourne and fuelled the drive to restrict access to the catchments completely. A policy of closed catchments had begun, which was to dominate the institutions in charge of Melbourne's water supply for decades to come (Plate 9).

The closed catchment policy was advanced by engineers and staff within the Water Supply Board. (The organisations responsible for the supply of Melbourne's water, and disposal of its sewage, have undergone numerous name changes during their history, including the Commission of Sewerage and Water Supply, Water Supply Branch of the Public Works Department, the Water Supply Board, Melbourne and Metropolitan Board of Works and, most recently, Melbourne Water.) Other supporters included scientific luminaries like Ferdinand von Mueller (1825–1896), the first government botanist in Victoria. There were many benefits to excluding all human activities from these areas. Aside from issues of public health, there was also the widespread belief among many scientists of the time that forests attracted rain, as well as filtering and cleaning water. This notion of forests attracting rain is an ancient one, and may have its origins in the writings of the Roman architect and engineer, Marcus Vitruvius Pollio (c80–c20 BCE). He argued that streams are best sought in mountainous areas

DAMNED IF YOU DO AND DAMNED IF YOU DON'T

The damming of a river causes significant alterations to the flow patterns downstream of the dam. In the absence of the usual seasonal flows of water, patterns of flooding and temperature changes, some fish and invertebrate species have disappeared from dammed rivers altogether. Some fish rely on increased winter flows to begin breeding. Others inhabiting billabongs alongside rivers may breed only in rare flood years, which become even rarer or non existent once dams are built. In some species warm summer water temperatures trigger breeding, but dam engineers typically release water from the cold bottom waters, even during summer.

When Museum Victoria researchers were asked to examine the effects of a new dam on the Thomson River, 140 kilometres east of Melbourne, they found that the ecology of the streambed below the dam was significantly altered. Increased sedimentation during construction reduced the diversity of aquatic invertebrates (Plate 8) in the area. Once the dam was closed, the water flow was reduced and the organisms living below the dam were quite different from those on undammed rivers.

Initially it was though that these differences were due to changes to water flow, with colder water released at different times of year than would occur naturally. However, a recent survey of many large dams across south-east Australia revealed another story. It seems that, no matter what type of flow regime is employed, no matter how old the dam or how natural the river beneath, certain species of invertebrate remain rarer than expected immediately below a dam. But this difference only extends as far as the next branch in the river. It seems dams are simply a physical barrier to the dispersal of many aquatic insects and their larvae, preventing them from drifting downstream. Fortunately, all it takes is another tributary (as a source for the dispersing organisms) to restore the balance.

because, in those situations, they are generally sweeter, more wholesome, and more copious, on account of their being sheltered from the rays of the sun, of the trees and shrubs in those places being in greater abundance, and of the sun's rays coming obliquely on them, so that the moisture is not carried off.

The notion that forests could actually increase local rainfall and conserve water for stream flow during drought was reinforced by observations that forests were cooler and more humid at their centres and that rain gauges in forest clearings collected more water than those in open areas. This latter observation, however, was probably due to the fact that the surrounding forest interferes with air currents that reduce the amount of water collected. Forests are often located at higher altitudes, where rainfall is naturally greater than in cleared, low lying regions. The perceived link between forests and water was a strong one, however, and dominated the textbooks on forest hydrology at the time.

Although Mueller supported the link between rainfall and forests as a major reason for the closed catchment policy, he also added a more passionate and ethical dimension to his support for the forests. He argued that the forests were 'a heritage given to us by nature, not for spoil or to devastate, but to be wisely used, reverently honoured and carefully maintained'. Like many other early Australian biologists, Mueller was a pioneering conservationist, and his actions contributed directly to the preservation of many of the forests we value so highly today. But more obviously, it was the actions of the Water Supply Board, and particularly its engineer, William Davidson (1844–1920), which ensured the success of the closed-catchment policy. Davidson tirelessly and vigilantly protested against plans to subdivide and settle, not only in catchment areas then used for metropolitan water supplies, but also in areas with the potential for future use. This remarkable foresight ensured the protection of vast areas of forest which are still sufficient to provide a clean supply of water to the three million Melburnians of today.

By the late 1800s (just before its banking collapse in 1892) Melbourne was one of the richest cities in the world. With a population of half a million, it had more citizens than much older cities like Birmingham or Boston. It was a characteristically new city, in which a house with a garden was the ideal, and it consequently occupied a large area. At the time only London was 'bigger'. In Melbourne a mere 23 people occupied each acre, and even in the most crowded inner city suburbs densities only reached 37 people per acre – a far cry from the 500 who crowded into every acre of London's most densely populated neighbourhoods. An

WHEN NO MEANS NO

Despite the closed-catchment policy, Melbourne's catchments have faced continued environmental hazards. The forestry industry had for many years cast avaricious glances at the heavily timbered closed catchments. Indeed, from the earliest days timber getters were staunch opponents of the policy. With the Water Board's exceptional record of fire control (aided, no doubt, by the ready supply of water), these forests had matured into some of the finest sources of timber in the region. Foresters despaired as they saw mature specimens of Mountain Ash (*Eucalyptus regnans*) decline into old age, to be 'wasted' as they finally fell to the forest floor and decayed.

Forestry pressure increased in the 1950s, when the old link between forest and rainfall was finally dispelled by scientific research. Studies in North America even demonstrated that cleared lands produced a greater run off of water than forested regions. Foresters leapt at the opportunity to do the Water Board an enormous favour by harvesting its catchments! However, the Water Board was not about to abandon its longheld policy just because of research conducted on completely different ecosystems. It began its own series of detailed studies on controlled regions under different catchment use regimes. This research clearly demonstrated that the water yield of Mountain Ash forests is at its maximum when the trees are over 100 years in age. Young trees, such as regrowth after fire or logging, use up to 50 per cent more water than older trees. Older trees shelter more of their surface area beneath the canopy and therefore transpire less moisture and draw less moisture up from the ground. Vitruvius was on the right track, 2000 years ago, when he thought that the forest canopy protected water from evaporation. Removing the forests completely is not an option, since the resulting erosion would dramatically reduce water quality.

extensive suburban train and tram network (which, remarkably, survives to this day) enabled workers to live far from their place of work, facilitating the suburban spread. The provision of water from the Yan Yean reservoir allowed most homes in Melbourne to boast a bathroom at a time when most houses in Britain could not. As capital of the much lauded 'garden state', Melburnians had an obligation to maintain their gardens in a state worthy of cool climate ideals, despite inevitable droughts and hot, dry summers peaking at over 40 degrees C. Much of Melbourne's water supply poured onto the ground in one form or another.

By 1890 the average daily water consumption was 228 litres (50 gallons) per person or over 100 million litres (24 million gallons) in total. The Yan Yean water supply had been joined by another aqueduct drawing water from further up the Yarra River on Watts River. This level of supply was exceptionally high, even by international standards. Few British cities could deliver even half this amount. But the ready supply of water came at a cost. The ad hoc local drainage systems could not cope with the extra supply. Most houses simply drained their sullage water into a 'soak' in the backyard. Lucky suburbs were able to drain their surplus water into the lanes and back alleys from which the 'night men' collected their odorous buckets of excrement. At best, the grey water found its way into the local creeks and tributaries of the Yarra (as did some of the night men's collections). At worst it formed great stinking cesspools of rank, disease-ridden sludge. Turn of the century Melbourne was constipated.

Deaths from typhoid were rising in Melbourne, while they were declining in other cities. Both a clean water supply (which Melbourne had thanks to ever-expanding closed catchments) and underground sewers

Engineers' drawings for types of house connections for sewerage in Melbourne, History and Technology Collection, Spotswood, Museum Victoria.

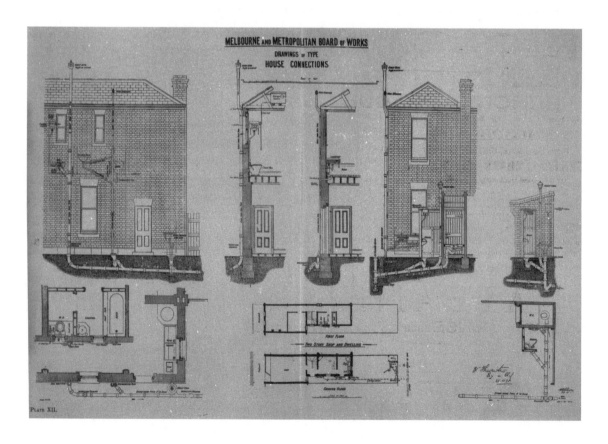

Part 2 INTO THE FORESTS

were recognised as important components in the battle against typhoid. Deaths from typhoid were much lower in Melbourne's sister cities of Sydney (which was partially sewered) and Adelaide (which was fully sewered). Despite Melbourne's reputation as a burgeoning modern city, typhoid became its special shame. An organised sewerage system was inevitable.

Sewerage, however, was a contentious issue: few municipal authorities were prepared to co-operate sufficiently to arrange a co-ordinated service. Most of the small townships surrounding Melbourne (which now constitute inner suburbs) were located either on the Yarra River or its tributaries, Merri Creek, Moonee Ponds Creek and the Maribyrnong (or Saltwater) River, and these waterways provided 'free' drainage. But nature has a way of making it clear when she has had enough. Just as London suffered its 'Great Stink', Melbourne became known as 'Smellboom' or 'Smellbourne' as the city soaked in filthy water. The Scottish traveller James Goudie had regarded the Liffey in Dublin to be the foulest river in Britain and Ireland, but found that to be sweet compared to the Yarra, which he ranked as 'the filthiest piece of water I ever had the misfortune to be afloat on'.

Even once agreement was reached that sewerage was essential, money continued to be a source of delay. Melbourne's gold fuelled boom ended in a dramatic economic crash in which many of Melbourne's banks closed their doors; much of the money devoted to funding a sewerage program was lost or out of reach. Overseas loans, long a favoured source of income for such schemes, were no longer available. But the sewerage program went ahead anyway and, ironically, played an important role in re-establishing Melbourne's shaky local economy. Bonds floated on the local stock exchange were seen as a safe investment in Melbourne's future, restoring confidence in an uncertain economic environment. The massive public works program required to build a sewerage system for the entire city provided a critical boost to Melbourne's economy.

Once Melbourne was committed to a sewerage system, debate revolved around the type of system to be built. Some early proponents favoured an ocean outfall, which would remove any possible unpleasant odours and eliminate the prospect of Melbourne being held to ransom by irate sewerage workers. Fortunately officials decided to risk disgruntled sewerage workers and opted for a sewage farm at Werribee, on the western side of Port Phillip Bay. The 'reuse' of sewage for agricultural purposes was a popular concept in the English agricultural debates of the time. Pumping of sewage into rivers and oceans was seen as wasteful when the increasing intensification of agriculture demanded an ever greater use

Sewage was gravity-fed through a complex system of underground tunnels to its lowest point at the Pumping Station in Spotswood, shown here. Massive steam pumps forced the sewage upward until it was high enough to drain down along the Main Outfall Sewer to Werribee. The Pumping Station operated from its opening in 1897 until the 1950s, when it was replaced by a larger station at Brooklyn. The Pumping Station at Spotswood is now the historic centrepiece of Museum Victoria's technology campus, Scienceworks.

of artificial (and expensive) fertilisers. In any case, Port Phillip was not sufficiently tidal to allow untreated sewage to be deposited directly into it, and Melbourne was too far from the ocean to seriously contemplate an ocean outfall. Werribee Farm (or the Metropolitan Farm) was, happily, both the most environmentally sustainable and the cheapest option.

Processing at the sewage farm operated under the principles of sedimentation and filtering. Effluent was pumped over empty paddocks or paddocks planted with Italian ryegrass, through which organic material was absorbed and settled into the soil. The grasses used up nutrients, and the sediments were filtered out through the soil. The cleansed water soaked into drainage channels before passing into the bay. In order to prevent the grasses from becoming old and rank, and to maximise the uptake of nutrients, mowing was required – a task efficiently performed by huge herds of cattle, making the sewage farm the largest beef producer in Victoria for many years. The simple system of grass filtration worked well, despite initial setbacks. For example, in winter, when rainfall is higher and evaporation lower, the farm became overloaded, requiring additional land filtration methods. Expansion and a system of holding tanks greatly alleviated this problem. In general, the system was efficient, despite many complaints about the smell. Some even claimed the farm could be smelt from the opposite side of the bay on a fine summer's evening. Proponents of the sewage farm protested that this particular smell

PLATE 8. Larvae of aquatic insects commonly found in Victorian rivers.
P. Marchant

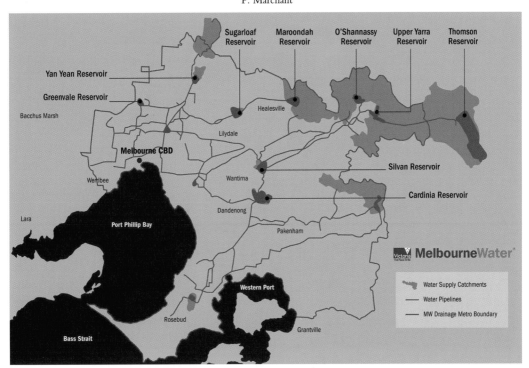

PLATE 9. Melbourne's current water catchments.
Melbourne Water.

PLATE 10. A Spotted Wobbegong Shark (*Orectolobus maculatus*) and Broadnosed Sevengill Shark
(*Notorynchus cepedianus*) collected from Port Phillip Bay in the 1800s.
F. Schoenfeld for F. McCoy, *Prodromus of Zoology of Victoria* (1878–90).

PLATE 11. 'Ferntree Gully in the Dandenong Ranges', by Eugene von Guérard.
National Gallery of Australia, Canberra.

was that of rotting seaweed on the beach at Werribee, not the farm itself. The seaweed along the shore near the Werribee outfall is, admittedly, particularly luxuriant. Despite its treatment, the sewage outflow remains high in nutrients and minerals.

The effect of sewage outfall on the bay has been an ongoing concern, particularly in the 1960s when it was discovered that heavy metals had risen to dangerous levels. Although the sewage farm was often blamed for this, factories in Corio and Hobson's Bay were discharging pollutants directly into the creeks feeding the bay and many used the untreated stormwater systems for waste disposal. Many contaminants are better dealt with via the sewerage works, where heavy metals settle out before they can enter the bay. Only control of what entered the system could halt the contamination of the bay, and gradually a system of tracing the sources of noxious wastes and prosecuting those responsible was developed.

The Yarra itself flows into the top of Port Phillip Bay, a large shallow body of water about 50 kilometres across with an average depth of only 12 metres. Other smaller rivers also flow into the bay, contributing their runoff to the 1940 square kilometres of water. This fresh water is countered by the influence of the tides which push through the narrow channel at the mouth of the bay. Water replacement in the bay is slow (about once a year) and its depth results in considerable turbulence and stirring of bottom sediments. The health of the bay was of as much concern as the health of the Yarra, particularly for residents of the wealthy beachside suburbs and for the local commercial fishery.

Unfortunately, the system of closed catchments which protects Melbourne's water supply from contamination cannot be applied to the river systems once they enter the metropolitan area. Like any major city, Melbourne produces vast amounts of waste which is not disposed of through regular networks. Because the city is built around the catchment system of a major river, much of this incidental waste ends up in the river system. The city itself, its roads and roofs, forms a hard impenetrable shield over the surface of the land, preventing water from soaking into the ground and dramatically increasing the amount of stormwater. Stormwater is the runoff from a million houses and factories, sluicing through kilometres of streets and gutters, collecting litter and dirt to deposit in the same creeks and rivers which have been washing away Melbourne's grime for the last 150 years. Rubbish in the street, soapy water from washing cars, toxic residue from washing paintbrushes, pesticide runoff from gardens, all end up in local rivers and creeks, many of which were converted to massive concrete drains to control flooding in lowlying areas.

THE HEALTH
OF THE BAY

The health and cleanliness of Port Phillip Bay has been a preoccupation for Melburnians for many years. Proposals in the 1960s for a new sewage discharge point on the east side of the bay prompted demands for a full environmental study into the health of the bay. Studies over the previous century formed the baseline for this research. In the late 1800s the Royal Society of Victoria collected plants and animals from the bay (which were lodged at Museum Victoria; Plate 10). The museum followed up this research in the 1960s with the first systematic survey of the wildlife of the bay. These collections form important biodiversity records against which to measure changes in the future. Similarly, recordings of salinity, nutrient flow, dissolved oxygen and nitrate levels made in the 1950s provide the basis for comparison in later years.

The full scale environmental survey in the 1990s was initiated by plans for a new sewage works. Museum biologists were joined by scientists working on the physics and chemistry of the bay. Surprisingly, the bay turned out to be much cleaner and healthier than anyone expected. The rich flush of algal growth below the Werribee outfall consisted of a diverse range of species, not at all indicative of pollution (which tends to cause a single group of algae to dominate). The communities of invertebrates inhabiting the sediment around the outfalls did differ from those nearby, but only in that they contained species which were more tolerant of fresh water (as would happen at a river outflow). Although the increased nutrients that the Werribee outfall adds to the bay do cause changes to its ecology, these changes are localised and do not significantly affect the functioning of the bay as a whole. Critical indicators of water quality (like levels of dissolved oxygen and phosphorus-to-nitrogen ratios) were favourable, and levels of toxicants were falling, in line with increasing restrictions on their flow into the bay. The report did, however, discover one major threat to the health of the bay indirectly linked to pollution. Exotic species, like Sabella and Pacific Sea Star, may have been introduced by ships emptying ballast water into the bay in preparation for loading.

Like any city built on a river system, Melbourne has long had a problem with floods. And like many boomtowns, Melbourne lacked early planning – suburbs were built on floodplains, disregarding natural drainage. The irregular, aseasonal cycles of Australia's rainfall exacerbate the problem. Estimates based on water discharges in previous years have proven drastically wrong. Prior to the droughts of 1915, Melbourne's existing water supply was thought to be adequate for a one third expansion of the city. After these droughts it was realised that the existing water supply was barely sufficient for the current population. Similarly, the floods of 1934 caused widespread devastation which no one could have foreseen. High tides and heavy rainfall pushed the coastline of neighbouring Westernport Bay inland by 24 kilometres to the north. One of Melbourne's main city streets, Elizabeth Street, became a roaring torrent as it reverted to its original, and unrespected, role as a part time tributary of the Yarra. Entire suburbs were inundated, some only visible as a forest of chimneys, as local creeks broke their banks and reclaimed their floodplains.

Flood control was clearly essential. Runoff from the hard city streets, unable to soak into the ground, had increased the peak flows of the creeks during times of high rainfall. Instead of draining away slowly and evaporating, the flood waters poured off the impervious artificial surfaces in an almighty rush. Despite arguments (presumably from those who lived in the wealthier, hill top suburbs) that the low lying suburbs should be abandoned to their fate, flood control measures were introduced. Many creeks were converted to concrete, or bluestone lined, drains which allowed massive quantities of water to move safely through the suburbs to the bay. But in their undue haste to shepherd the threatening waters away,

Yarra Falls Mill and
factories under floodwater,
1 December 1934.

the flood controllers also hastened the movement of sediments and litter into the bay. Without meandering creeks and riverside grass beds to slow the water down, there is no opportunity to deposit sediments along the way – the natural filtering and cleansing processes of the rivers have been lost. And of course, a host of natural wildlife based around the creeks disappeared as well.

But this is not an irreversible change. Increasingly, local councils and community groups are restoring their local creeks, replacing concrete channels with natural meandering creeks, wide flood plains and flood retarding basins to cope with sudden inundations. Litter traps help to remove some of the refuse from the system before it reaches open water, and residents are encouraged to reduce the load on local creeks by washing their cars on the lawn instead of the street and not treating their drains as toxic waste dumps. The results have been spectacular at a local level. Not only have some stretches of local creeks become pleasant parklands for humans to enjoy, but local wildlife has also given its vote of confidence by returning in abundance. Sacred Kingfishers (*Halcyon sancta*) return each year to breed along the once degraded Merri Creek while Platypuses (*Ornithorhynchus anatinus*) can be found in heavily built up regions (although, with their secretive habits, they are not easy to see). Perhaps, in time, the consequences of such healthy activity will flow on to the bay with decreased levels of litter and sediment, and the story of Melbourne water will be as green as that of any city, from protected forest catchments through local waterway parks and into a clean bay and open ocean.

5 FORESTS
OF FIRE

The forest-dwelling Coventry's skink (*Niveoscincus conventryi*) was named after the long-serving
Curator of Herpetology at Museum Victoria, John Coventry.
Peter Robertson.

OLD-GROWTH FORESTS are Nature's cathedrals. The ancient, moss-covered trunks of these forests can inspire near-religious awe. The roots of the vast trees form the forest's foundations and their trunks buttress the vaulted canopy of leaves, filtering light like monochromatic stained glass. Perhaps these aged trees suggest a longevity and serenity lacking in the frantic, time-strapped pace of modern human existence. Age and lack of disturbance are key factors in the definition of an old-growth forest. People like forests to look old and untouched by human hands – such forests are deemed to be 'natural'. But the attributes people use to designate a plant community as old or undisturbed do not necessarily relate to age or virginity.

People value 'old' forests and intuitively measure age in relation to the size of the trees. But not all old trees are big. Comparatively stunted tree communities such as scrub and low woodland are commonly rated as less 'natural' than forests, even though they may be as old as, or older than, their more grandiose cousins. In fact, many of the features we like in 'old-growth' forests say more about our own tastes in landscapes than they do about the integrity or age of the ecosystem they occur in.

For example, most people prefer forests with an open understorey that allows easy access. In truth, most people favour 'park-like' forests – a preference more readily confessed by early explorers than by contemporary admirers of scenic beauty. In 1826, the French captain Dumont d'Urville described a pristine area now covered by Melbourne suburbs and farms as 'a lovely grassland … shaded by fine trees … and looking rather like our royal forest around Paris.' Some people regard this preference as part of a colonial European inheritance; others argue that it is cross-cultural and reflects our evolutionary heritage in climbing down from the trees to utilise the grassy savannahs (although I have yet to see a study of Inuit landscape preferences to clarify this argument). Whatever the cause, most people regard a lightly cleared forest as more natural and less disturbed than an untouched forest with dense impenetrable undergrowth. Some levels of disturbance may be instinctively attractive, even though at a conscious level we tend to reject disturbance as unnatural and undesirable. But in fact, natural disturbances play a key role in maintaining these forests, not only in a state humans find attractive, but in the full glory of their ecological complexity.

The popular concept of 'old-growth' may be based on aesthetics, but the scientific concept of 'old-growth' is based on a linear, sequential model for plant communities originally developed for North American hardwood forests. Botanists traditionally described forest communities as developing from a succession of short-lived, unstable grass-, heath- and woodlands to a stable, self-perpetuating forest of high diversity. But most Australian forests, and other plant communities, are neither stable nor self-perpetuating. Rather, they are cyclical and diverse, constantly changing in structure, form and age as determined by natural disturbances like fire, flood and wind. The stability of Australian forests lies not in their longevity or self-sustenance, but in the maintenance of a rich mosaic of successional stages which supports a much greater diversity of plants and animals than an unchanging forest ever could. And nowhere is this better illustrated than in the heartland of what many people regard as 'old-growth' forests – the great Mountain Ash (*Eucalyptus regnans*) forests of central Victoria.

Beyond the ever-expanding eastern margin of Melbourne, a highway passes from densely packed city streets to fringeland suburbia. Half-built developments, light industry and pockets of farmland link the surrounding satellite towns into a metropolitan sprawl. The footprint of humanity is indelibly stamped on this landscape. But just twenty minutes from the suburban edge of Australia's second-largest city, the traveller is unexpectedly engulfed in the cool, venerable quiet of the southern temperate forests.

The hilly ranges to the north-east of Melbourne form the heartland of the great Mountain Ash forests. In winter, the highest peaks are dusted with snow, while their lower slopes harbour vestiges of the ancient temperate rainforests that once covered a wetter Australian continent. Dominating these forests are the enormous Mountain Ash. Towering over 100 metres high in places, Mountain Ash were once considered the tallest trees in the world. In 1881, to the east of the remaining forests, the Thorpdale tree was officially measured at 114 metres (375 feet) high – before being felled. A commemorative pole in a deforested landscape now celebrates the site of 'the world's largest tree'. Tales abound of taller trees whose demise was not officially recorded. No other hardwood grows as tall or as fast as the Mountain Ash, and its height is equalled only by a softwood (the North American redwood *Sequoia sempervirens*). But in addition to its prodigious height and speed of growth, the Mountain Ash produces strong, fine, flawless timber. While colonial carpenters despaired of the recalcitrant native hardwoods which blunted and broke their tools, the Mountain Ash proved a more pliable subject. To this day, shipwrights cannot enter the forest without seeing ships' masts, so smooth and straight are the enormous trunks of the Mountain Ash. Despite this fatal attraction, many old giants remain in the inaccessible gullies of the Acheron Way, surrounded by legions of upright young sentinels maturing to take their place.

Mountain Ash
(*Eucalyptus regnans*).
S. Madder.

Mountain Ash forests vary enormously, from uniform stands of young trees with a grassy understorey to mixed forests with a thick rainforest mid-canopy. Along the Acheron Way, Mountain Ash dominate the upper canopy and occasionally share the sun with smaller Messmates (*Eucalyptus obliqua*) and Mountain Grey Gums (*E. cypellocarpa*). In the cool, moist gullies, ancient rainforest trees occur – Myrtle Beech (*Nothofagus cunninghamii*) and Southern Sassafras (*Atherosperma moschatum*). These trees are the structural supports of the forest ecosystem. Beneath their vast shoulders shelter a wealth of gentler shade-loving plants: the prehistoric

A Mountain Ash
community.
S. Madder.

Part 2 INTO THE FORESTS

ferns, delicate epiphytes, mosses and lichens. Fragile webs of fungi pigment the decomposing remnants of the past. Antediluvian tree ferns (*Dicksonia antarctica*) create a human dimension to the arboreal cathedral, their bracken-like foliage forming an artificial ceiling a couple of metres above the ground. The rare starry Long-leaf Wax-flower (*Eriostemon myoporoides*) occurs only in the undergrowth of high country forests like these. The small and elegant Tree Geebung (*Persoonia arborea*) thrives in these forests but is rarely found outside them. Over 250 different plant species co-exist here, some stretching ever upward to spread their canopies in the sun, while others shelter in the cool, dank micro-climate created by their taller brethren (Plate 11).

These forests are home to both the famous and the fascinating. Rare Leadbeater's Possums (*Gymnobelideus leadbeateri*) dart in the moonlight, in constant fear of formidable Powerful Owls (*Ninox strenua*). Chocolate Wattled Bats (*Chalinolobus monio*) peer from a prized roosting site in a hollow tree, a home in high demand not only from other bats but also parrots, owls, possums and gliders. Thirty mammal species make their homes in these forests. More than sixty bird species divide their time between canopy and forest floor like inhabitants of a high-rise apartment block. Frogs, snakes, lizards and fish add a further thirty-five vertebrate species to the list.

The abundance of vertebrates pales to insignificance when compared with the variety of invertebrate life housed in the forest. Hundreds of species of terrestrial arthropods inhabit every nook and cranny. The hefty green larvae of the Emperor Gum Moth (*Opodiphthera eucalypti*; Plate 12) munch contentedly on their namesakes. Numerous species of tiny foliar mites eke out their entire existence clinging precariously to the leaves of the upper canopy. The leggy velvet-worm *Peripatoides*, descendant of a venerable lineage 500 million years old, perambulates through the leaf litter in the company of innumerable earthworms and leeches. The bed of the Acheron River itself is home to at least one hundred different species of aquatic insects, crustaceans and other invertebrates.

But the most abundant form of life in this forest is not visible to the naked human eye. Inhabiting the soil, the water, the plants, the animals and even the rocks are countless legions of microscopic single-celled organisms. No-one knows how diverse or abundant these creatures are. There may be thousands of different species of soil bacteria, all playing an indispensable role in the forest's survival. These invisible armies (along with other detrivores such as worms, fungi, beetles and flies) break down the carcasses of other creatures. They return precious nutrients to the soil and release them for re-use by future generations.

The complexity and vast size of the Mountain Ash forests suggests that they are classic old-growth forests that have reached the pinnacle of their development. At first sight, a linear, sequential succession process, culminating in a stable, self-perpetuating forest, appears to be perfectly appropriate to describe the development of a Mountain Ash forest. Indeed, early settlers in the region were astonished at the seemingly indomitable powers of rejuvenation displayed by the Mountain Ash forests reclaiming arduously cleared land.

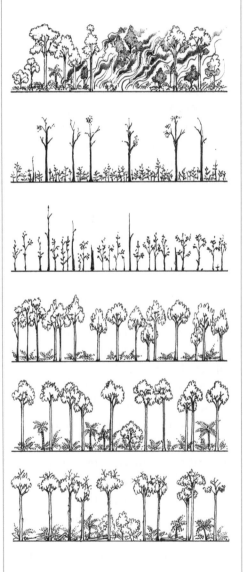

The Mountain Ash is the phoenix of the plant kingdom: this giant of the forest is reborn in the crucible of fire. Mature Mountain Ash shed most of their seeds on the remains of their funeral pyre. The death of the mature tree causes its seed cases to dry and open, an effect exacerbated by the heat of the fire. In the blackened aftermath of an intense forest fire, myriads of tiny seedlings shoot up in the ash-enriched, clear soil to form a dense, young forest of upright saplings. Only when the forest is all but destroyed are the conditions right for Mountain Ash to reproduce.
S. Madder.

After clearing, only grasses and ground-covers, such as the fireweeds (*Senecio* spp.) and Giant Mountain Grass (*Dryapoa dives*), appear. But even these short-lived, rapidly spreading plants are outpaced by the phenomenal burst of speed put on by the young Mountain Ash saplings. Millions of tiny seedlings emerge from the ground in a mad race for survival and, within just ten years, the once bare ground is a vast thicket of young trees, each within metres of their neighbours. Their roots form a thick, impenetrable mat, excluding other plants from taking hold. Battling against the regenerating forest proved too much for many would-be farmers whose lands were soon overtaking by a seemingly endless supply of young Mountain Ash. But Mountain Ash are extremely intolerant of shade. As they grow, the trees begin to thin out. The tallest and strongest of the young Mountain Ash trees cast their canopies with fatal effect over their smaller rivals. In the resulting gaps, more shade-tolerant species like wattles and smaller eucalypts emerge. In moist areas, tree ferns appear and in the now dark and humid understorey, ferns and mosses find a sheltered home. The remaining Mountain Ash, thinned by competition from two and half million seedlings per hectare to a mere 20 trees per hectare, grow larger and larger to take their place as the giants of an 'old-growth' forest.

Mountain Ash forests continue to thrive for hundreds of years if they are undisturbed. But at about 120 years, the Mountain Ash trees lose their youthful vigour. Over time, their tall, sparse canopy thins and becomes patchy. They lose limbs and their rotten stumps hollow out to make homes for gliders and parrots. Eventually they decay and fall, crashing to the ground where they slowly rot into the rich earth below. But on the shady forest floor, there is no place for the sun-loving young Mountain Ash to grow. On cool damp sites rainforest species proliferate, while on the drier slopes low shrubby vegetation prevails. The Mountain Ash forests simply disappear, unable to compete successfully for light under an established canopy. This may be an old-growth forest, but it is neither stable nor self-perpetuating. Its survival is dependent on its own destruction – by devastatingly intense bushfires.

Strange though the Mountain Ash story seems, it is typical of many Australian ecosystems. Fire, usually regarded as a destructive force, is a necessity for many Australian plants. Rather than forming a stable 'old-growth' community of plants, many Australian plant communities engage in a regular and violent cycle of destruction and rebirth. Just as the young forest is part of an ecological progression to an old forest, the old forest is part of the natural progression to a young forest – with the assistance of fire.

A HUMAN FIRE HISTORY

Fires have become more frequent in Australia over the last 100,000 million years or more. Traces of charcoal and pollen in lake sediments near Canberra provide us with a detailed chronology of the region's fire and vegetation history. Lightning strikes probably caused most of these early fires, with such conflagrations becoming more common as Australia became drier and hotter. Fires also became more frequent when Aboriginal people began firestick burning more than 40,000 years ago.

Early European observers frequently noted smoke and fire-blackened areas along the Australian coast. They also found open, grassy woodlands that made travel easy and promised good grazing. We now know that Aboriginal people have used fire for millennia to promote a rich mosaic of environments. Firestick burning produces a flush of rapid growth favoured by grazing animals. It opens up the country for travel. Fire favours plants with underground tubers, like the Yam Daisy or Murnong (*Microseris scapigera*) which formed the staple diet of many Victorian Aboriginal communities. The Yakka or Grass-Tree (*Xanthorrhoea* spp.; Plate 13), which is neither grass nor tree, but a striking primitive plant with a thicket of grass-like leaves atop a tall blackened stump, produces edible flowers after fire. So fire-dependent are these plants that the home gardener is advised to greet newly acquired grass-trees with a welcoming blow-torch to settle them in to their new environments. Indigenous people in central and northern Australia continue to use fire to 'look after country'.

European Australians more commonly experience fire as a catastrophe that periodically sweeps the country. These are regular, yet devastating, events. Each summer is punctuated with a sense of dread about the impending fires and their possible severity. The worst, in south-eastern Australia, are known by name: Black Thursday, 6 February 1851; Red Tuesday, 1 February 1898; Black Sunday, 14 February 1926; Black Friday, 13 January 1939; and Ash Wednesday, 16 February 1983. On Black Friday, for example, 10 per cent (1.4 million hectares) of Victoria was in flames and almost all of its mountain forests were lost. The majority of today's Mountain Ash forests count their birthday from 1939.

Although bushfires were devastating and destructive, even the early settlers seemed to recognise the inevitability of fire in

Australia. While countries like the United States embarked upon ambitious programs to dominate nature and eradicate 'wildfires', Australian fire-fighting authorities employed a damage-limitation model which used fire to fight fire. While Americans developed 'water-bombers' to suppress fires, Australians used 'fire-bombers' for backburning or for control burns during low-risk conditions. American fire-fighting authorities have only recently acknowledged the necessity of fire in maintaining natural ecosystems (and preventing catastrophic wildfires) and have begun to deploy tactics like those developed in Australia to limit their severity. In Australia, fire has always been an inescapable, inextinguishable force of nature – inexorably burnt into the landscape.

Australia's forest history has been an ongoing struggle between rainforest and eucalypt. While elements of each coexist within the other, the dominance of each ecotype waxed and waned under changing climatic conditions.
S. Madder.

Australia is one of the most fire-prone countries in the world. No other continent has undergone pyric purging with such frequency over such a long period of time. Only in Australia has fire had such a major effect on vegetation over such a large area. Almost half of all Australian trees and shrubs belong to just two families – the *Acacias* (or wattles) and the *Myrtaceae* (which includes the eucalypts, tea-trees and bottlebrushes). It is the characteristics of these two dominant families that typify the Australian bush. During his midsummer visit to Australia in January 1836, Charles Darwin (1809–1882) described how the

> extreme uniformity of the vegetation is the most remarkable feature in the landscape of the greater part of New South Wales. Everywhere we have an open woodland, the ground being partially covered with a very thin pasture, with little appearance of verdure. The trees nearly all belong to one family, and mostly have their leaves placed in a vertical, instead of, as in Europe, in a nearly horizontal position: the foliage is scanty, and of a peculiar pale green tint, without any gloss. Hence the woods appear light and shadowless.

But it is smell, rather than colour, which creates the most distinctive impression of the Australian bush. As the antagonistic expatriate in Henry Lawson's 'His country – after all' discovers, the scent of gum trees is both unmistakeably and powerfully evocative of the Australian landscape – particularly when they burn. 'There was a rabbit trapper's camp among those trees; he had made a fire to boil his billy with gum-leaves and twigs, and it was the scent of that fire which interested the exile's nose, and brought a wave of memories with it.' The characteristically crisp aroma comes from the volatile oils of the *Myrtaceae*, which are distilled from the most pungent species and bottled for medicine cabinets around the world as 'eucalyptus' and 'tea-tree oil'. The role of these volatile oils, however, extends far beyond eliciting nostalgic reminiscences or treating minor ailments and provides us with an olfactory clue as to the nature and origin of many Australian forests.

Over the past 250,000 years Australia's climate has become drier and hotter as the continent inches northward. Outside the ever-decreasing rainforest pockets, only plants that can cope with heat and drought survive. Aridity has favoured the evolution of 'sclerophyllous' vegetation (from the Greek *skleros*, hard, and *phullon*, leaf). Thick, leathery leaves do not need to be regrown very often. Waxes and oils replace thinner fluids which evaporate too rapidly. The unusually heavy toll of insect grazers on Australian plants probably also selected for unpalatable chemical compounds and tough leaves. These tough, oily leaves and the woody seedcases associated with them pre-adapted members of the *Acacia* and *Myrtaceae* families for surviving fire and enabled them to exploit a fire-prone environment.

FIRE-ADAPTED PLANTS

Fire-dependent plants often 'enhance' fires by producing large amounts of leaf and bark litter containing highly flammable, volatile oils. The Mountain Ash sheds great strips of bark down the length of its smooth trunk, creating a massive tangle of bark and leaves. The apt name of the Candlebark (*Eucalyptus rubida*) leaves no doubt as to the effect of its multi-hued shedded bark. Such kindling might encourage a fast, cool fire to kill off competitors and undergrowth while allowing the tree's own seeds to survive the blaze. Large sheets of burning, loose bark may also blow ahead of the fire and facilitate the spread of spot fires. Such seemingly self-destructive behaviour favours those species that can survive the flames or use fire as a reproductive trigger.

Many plants that are destroyed by fire set seeds to regenerate in their place. The bright orange Hairpin Banksias (*Banksia spinulosa* var. *cunninghamii*; Plate 14) retain seeds in woody fruits that open when the parent plant is burnt. The burnt ground makes an ideal seedbed that is temporarily cleared of shade, litter, parasitic fungi and seed-eating ants. The seeds of wattles, which develop from the familiar soft, golden flower-balls, can lie dormant in the soil for years, germinating in response to abrasion, heat or smoke. While the Mountain Ash uses this strategy over a time scale of centuries, some small, short-lived herbaceous plants apply it over the course of a few months. Such 'fireweeds' produce masses of far-dispersing seeds that rapidly colonise burnt areas from unburnt surrounding vegetation.

Not all fire-dependent species die by fire like the Mountain Ash. Many other plants are capable of regenerating after fire. In the foothills of the mountain forests, Stringybarks (*Eucalyptus baxteri* and *E. obliqua*) encase themselves in thick, fibrous, thermally-insulating bark to protect dormant buds. These buds can resprout if the canopy is destroyed by light fire. The twisted, swirled trunks of the Snow Gum (*E. pauciflora*) swell beneath the ground with food reserves to supply regrowth after fire. Such lignotubers are common to most eucalypts, but notably absent in the Mountain Ash and similar fire-sensitive species. Some grasses, such as spinifex and blady grass, also resprout from leaf bases, while bracken resprouts from subterranean stems.

Which species survive and proliferate depends on the frequency and severity of the fires. Too many fires can destroy seedlings before they are old enough to produce their own seeds. Too many fires destroy regrowth before reserves have been replenished. Too few fires may see the adult plants die of old age without reproducing. Too few fires may see one or two species outcompete their neighbours, reducing the diversity of the community. The intensity of the fires also determines the nature of the community that regenerates. Intense fires often favour a grassland by completely clearing the landscape. Low-intensity fires may merely change the forest understorey. An occasional 'cool' fire can exclude sensitive rainforest plants but allow fire-resistant trees to survive, perpetuating an open woodland. Both frequency and intensity determine the success of the regenerating Mountain Ash forests. These dense forests with heavy fuel loads need infrequent but intense fires.

One of the key advantages of fire is that it is sporadic and random, affecting different areas, at different times and at different intensities. Fire creates a mosaic pattern of vegetation across the landscape, providing a diversity of habitats, including the refugia of older species in cool moist gullies where fire rarely ventures. This range of habitats accommodates the varying needs of a diverse spectrum of species – cleared land for the sun-lovers; cool, shaded gullies for the fire-sensitive ferns and mosses; dappled understoreys for the shade-lovers; and high canopies for the epiphytes and parasites. Such a wide range of contiguous habitats promotes diversity among both plants and the animals they support – and this is particularly noticeable in the Mountain Ash forests.

The current Mountain Ash forest is relatively homogenous in age, about sixty years old. But in the past, when the forest was four times its current size and fires affected smaller portions at a time, a much greater diversity of ages may have existed within it. The age of the forest determines which habitat features are present and how suitable it is for certain species to use.

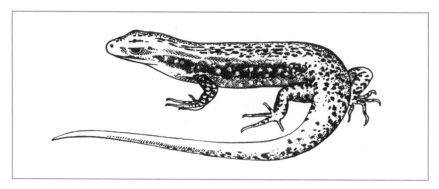

Southern Water Skink
(*Eulamprus tympanum*).
S. Madder.

Part 2 INTO THE FORESTS

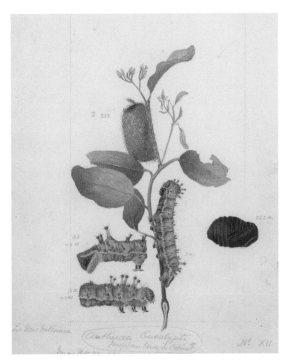

PLATE 12. The Emperor Gum Moth caterpillars (*Antheraea eucalypti*, now *Opodiphthera eucalypti*) illustrated by Arthur Bartholomew. Frederick McCoy, *Prodromus of the Zoology of Victoria* (1878–1890).

PLATE 13. The Austral Grass-tree (*Xanthorrhoea australis*) only reproduces after fire.

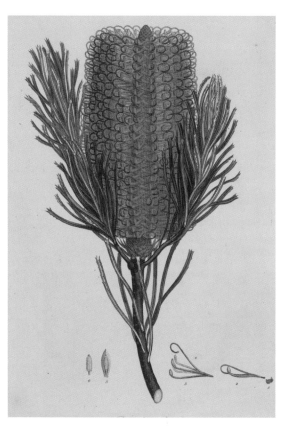

PLATE 14. Hairpin Banksia (*Banksia spinulosa*)
only germinate after fire in response to smoke.
By James Sowerby.

A Specimen of Botany from New Holland by James
Edward Smith (1793), State Library of Victoria.

PLATE 15. The Leadbeater's Possum (*Gymnobelideus
leadbeateri*) illustrated in a very uncharacteristic pose.
Frederick McCoy, *Prodromus of the Zoology of Victoria*
(1878–1890), vol. 9, Plate 81.

ANOTHER SKINK'S TALE

The Southern Water Skink (*Eulamprus tympanum*), pictured on page 68, is common to Mountain Ash forests. It basks in the sun on fallen logs and bare ground in both recently cleared areas and old forests where the canopy is patchy. Spencer's Skink (*Pseudemoia spenceri*) is another sun-lover; however, this highly arboreal species basks on trees up to 75 metres high in the canopy. Patches of sun under a broken mature-forest canopy suit it as well as sunlight on the exposed dead tree trunks.

McCoy's Skink (*Nannoscincus maccoyi*) draws heat not from the sun but from its surroundings. It lives in warm, moist, composting leaf litter. Unlike the Southern Water Skink and Spencer's Skink, this species is most abundant in the oldest, moistest parts of the forest. The recently discovered Coventry's Skink (*Niveoscincus coventryi* – named after Museum Victoria's former curator of herpetology, John Coventry) is also happy in shaded forests with leaf litter. Unlike McCoy's Skink, Coventry's Skink isn't dependent on the warm litter but maintains its temperature by shuttling between sun and shade. It can therefore be found in forest of any age with some litter and is the most common skink of young forests.

Imagine a timelapse image of a Mountain Ash forest from bare ground to maturity at 150 years old – from a skink's perspective. The first frames are dominated by sun-loving Southern Water Skinks on the ground and Spencer's Skink on the burnt tree stumps. Within a decade, the young forest has shaded out these species and begun to provide leaf litter for Coventry's Skink, which alone dominates the forest for the next fifty years. As the forest thins and matures, Southern Water Skinks and Spencer's Skinks recolonise, exploiting patches of sun gleaming through the broken canopy. Finally, in the deep moist litter of the very old forests, McCoy's Skinks appear.

In a diverse age forest, the skink populations rise and fall with the age of the stand; as one patch is cleared, the population falls while in another ageing stand it increases. The mosaic pattern of forest ages enables animals to move from area to area to exploit the particular age forests they require, and retreat to other areas when these forests are burnt or are no longer suitable.

The cool, calm forests east of Melbourne remain, like mature forests everywhere, beautiful, serene and inspiring. In an ecological context, however, they are not necessarily more valuable or important just because they conform to our subjective notion of an old-growth forest. They are not even particularly old. The often twisted and gnarled Huon Pines (*Dacrydium franklinii*) of Tasmania, for example, are a mere quarter the stature of a mature Mountain Ash, but number their lifespans in thousands of years. 'Old-growth' forests are no more nor less important to preserve than native grassland, heathland or alpine bog. Our attachment to 'old-growth' sometimes leads us to forget that preservation cannot involve the suspension of time and exclusion of any disturbances. Disturbances, like fire and storm-damage, are a natural and necessary feature of the survival of these ecosystems. What is important is that enough naturally vegetated areas are left so that fire can create a mosaic pattern of regeneration within each ecosystem, rather than devastating the last remaining patch.

A vast variety of species depend on a rich mosaic of habitat for their survival. Victoria's faunal emblem, the Leadbeater's Possum (*Gymnobelideus leadbeateri*), provides a potent example of a vulnerable species dependent upon a diversity of forest habitats. An optimal habitat for a Leadbeater's Possum has abundant old trees and a vigorous wattle understorey. Only trees older than 200 years provide the hollows that these possums need to shelter and breed. Silver Wattle (*Acacia dealbata*), Mountain Hickory Wattle (*A. obliquinervia*) and Forest Wattle (*A. frigiscens*) provide the Leadbeater's Possums with their staple winter diet of sticky, sugary gum on which they depend when insects are scarce. In the seventy years since the major 1939 fires, a perfect Leadbeater's heartland has developed in the Mountain Ash forests with a mature wattle understorey punctuated by hollowed dead trees left by the fires. But these conditions cannot last. The wattle understorey itself is not long-lived. The young Mountain Ash will not begin to develop hollows for another century and a half. And the old trees are already beginning to decay and collapse. The comparatively homogenous nature of the remaining Mountain Ash forests provides few alternative havens for the possums. Without the mosaic, fire-induced pattern of diversity within the forest, species like the Leadbeater's Possum may literally not have room to wait for the forest to regenerate.

6 THE MYSTERY OF THE REAPPEARING POSSUMS

A mounted specimen of Leadbeater's Possum found
by Baldwin Spencer in his accountant's office.
Mammology Collection, Museum Victoria.

EUROPEAN SETTLEMENT OF Australia had a dramatic and devastating impact on the native wildlife. Europeans have a very different heritage of land management to the indigenous inhabitants of Australia. They harvested timber and cleared the land of 'unproductive' vegetation – scrub, mallee, heath, old tree stumps. They planted cycles of annual crops, leaving the soil exposed to erosion. They drained swamps, irrigated dry lands and dammed rivers, thus destroying the breeding grounds of many species, raising ground salts to create salinated wastelands, and preventing life-giving flood waters from washing over the parched floodplains. They suppressed small, periodic fires for fear of loss of property, fences and stock. Without fire, the untamed undergrowth grew thick and unmanageable for the struggling pioneer farmer until, fuelled by years of unburnt kindling, massive, uncontrollable fires raged across vast areas, devastating human and animal communities.

And finally, Europeans colonised in company. They brought not only sheep, cows, pigs, goats, horses and camels but also other less useful creatures, either for company, for sport or by accident: rats, mice, rabbits, foxes, cats, dogs, starlings and sparrows. The hard-hoofed ungulates carved up the fragile topsoils of inland Australia. The rabbits did what rabbits do best, while eating native Australian animals (and introduced stock) out of house and home. The remaining native species shelter in remnant vegetation where they are fair game for a new range of predators: foxes, dogs and cats. Since Europeans arrived in Australia more than twenty terrestrial mammal species have disappeared, and another sixty are on the critical list. Not a bad record for a mere 200 years.

But simple explanations cannot always account for the complex circumstances surrounding extinctions, much as we like an easy scapegoat. It is easier to write off the dinosaurs with a massive meteorite strike than discuss the complex patterns of extinctions that occurred over millions of years long before and after the Chicxulub meteor hit the Earth. It may be simple to blame cats for the disappearance of the Greater Bilby (*Macrotis lagotis*) from the Adelaide Hills, but more disturbing to consider the interactions between mass habitat destruction, competition and predation. Despite the attractions of simplistic answers, ecological phenomena like extinction are rarely simple. Undoubtedly, the arrival of Europeans has had a devastating impact on the Australian landscape and its wildlife, but shifting climatic patterns and the vagaries of evolution ensure that some species were rare even before Europeans arrived. For them, changes to land-use patterns, fragmentation of habitats and the introduction of predators and competitors may be the last straws, rather than the primary cause of their decline. Life for any species, over the vastness of geological time, is pot-luck and filled with uncertainty. This is the story of two such species, hanging by a thread on the verge of extinction.

The Wombeyan Caves have been a popular picnic spot in the Blue Mountains south-west of Sydney since the early 1870s. Visitors then were provided with a candle and allowed to roam throughout the five-chambered cave complex. Not only were the caves themselves of interest, but the area surrounding them typifies the scenic beauty of the Blue Mountains region. Jagged red-yellow cliffs and ridges provide sweeping views across the forested gullies and valleys. On the drier slopes, eucalypts and wattles prevail, while in the cool dampness of the valley floors older rainforest remnants like Tree Ferns (*Dicksonia antarctica*) and Antarctic Beech (*Nothophagus moorei*) survive. A long winding mountain road leads the visitor down through the densely forested slopes, into a steep narrow valley to the caves themselves.

On 9 November 1894 one such visitor was exploring the ridges above the caves when he came across the exposed floor of a long-eroded cave. Small clumps of reddish-brown rock were flecked with tell-tale white fossil fragments. The visitor carefully collected a pile of these rocks and took them home.

This visitor was no ordinary picnicker taking home a memento of his day's exertions. The gentleman in question was a young Scottish doctor, Robert Broom from the nearby town of Taralga. Broom was an enthusiastic naturalist and made numerous contributions to the study of Australian mammals during his time here. Later in his career Broom became a renowned palaeontologist in South Africa. He was the first to discover the remains of ancient mammal-like reptiles whose descendants diverged into both modern reptiles and mammals. He also uncovered some of the earliest remains of human ancestors in the Transvaal. Even at this early stage in his career, Broom recognised an unusual find, and after carefully cleaning some of his collection, he believed he had found two new species of marsupial, probably possums.

Medically trained, Robert Broom (1866–1951) became the world's leading expert on the mammal-like reptiles of South Africa; he also published a major monograph establishing that australopithecines were hominids.

The fragments included tiny jawbones a little more than a centimetre in length. Many possums are hefty cat-size creatures, like the Common Brush-Tailed Possum (*Trichosurus vulpecula*), but these fragments belonged to animals so small they could curl comfortably in the palm of his hand. One of them Broom named *Burramys parvus*, which incorporates the Latin words for 'small' (*parvus*) and 'mouse' (*mys*) with the local Aboriginal name for the region, Burra Burra, or place of rocks. Broom could not have known at the time how apt 'small rock mouse' was for the fossil he had discovered. The second animal he termed *Palaeopetaurus elegans* (or elegant ancient possum).

Burramys excited much scientific interest, for although it seemed to be some kind of possum, its jaw contained high grooved cheek teeth previously regarded as characteristic of kangaroos. Was this an extremely small kangaroo? No-one knew, for no more fossil *Burramys* could be found in the caves and Dr Broom left Australia in 1896, taking with him the rest of his Wombeyan fossils.

Broom took most of his fossils back to Scotland and deposited them in the Anatomical Museum of the Medical Faculty at the University of Edinburgh. They remained there for over fifty years, until a colleague of Broom's asked a research student from Australia, David Ride, to re-examine them. Ride confirmed that *Burramys parvus* was in fact a member of the pygmy-possum family and related to the gliders, not the kangaroos. He also found that Broom's second discovery, *Palaeopetaurus elegans*, was not actually a new species, but rather the exceedingly rare Leadbeater's Possum, *Gymnobelideus leadbeateri*, which had not been seen alive since the early 1900s (Plate 15).

Leadbeater's Possum had been named by Frederick McCoy (1823–1899) at the National Museum of Victoria. He named it after John Leadbeater (1831–1889), the skilled taxidermist who prepared most of the museum's specimens. Two male possums had been collected in 1867 from the banks of the Bass River near Melbourne. McCoy believed this new animal was related to the Sugar Glider (*Petaurus breviceps*) but lacked the gliding membrane from wrist to ankle. Certainly, Leadbeater's Possum shared the Sugar Glider's petite proportions, silky grey coat and endearing striped face. But unlike the Sugar Glider, which is relatively abundant, no further trace could be found of the Leadbeater's Possum.

Burramys was known only as a fossil: there was no reason to think that it was still alive. But Leadbeater's Possums were known to have lived in recent times and so some effort went into finding new populations. Alas, these searches were to no avail. Three more specimens of Leadbeater's Possums did appear in the forty years after McCoy's discovery, in rather strange circumstances, but none were found alive in their native habitat.

Skull of a *Burramys parvus*, showing the unusual kangaroo-like tooth structure.

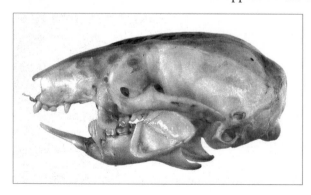

The third Bass River Leadbeater's Possum was donated to the museum in 1899 by a Mr A. Cole, although nothing is known of when he obtained the specimen. A few years later, Walter Baldwin Spencer (1800–1829) was visiting an accountant when he noticed an unusual possum mount. At first glance it

looked like a small squirrel, sitting upright on its haunches with front paws curled close to its chest and its tail fluffed out and curved in a characteristically squirrel-like 'S'. But it was not a squirrel – to the trained eye, it was clearly a Leadbeater's Possum. This possum had come from Koo-Wee-Rup Swamp, 65 kilometres north of Bass River where the original animals had been obtained. The possum eventually joined the earlier specimens in the museum collections.

Even at this early stage in the museum's history, the collection was a vast repository of material and not all of it was well-catalogued or even identified. Unlike today, where a number of specialist curators are employed for their expertise in identifying particular taxonomic groups, in Spencer's time there were no curators. It was not until a former preparator, Charles Brazenor (1897–1979), was appointed Curator of Mammals in 1931, that a fourth Leadbeater's Possum was discovered in the museum's own collection. The only information attached to the specimen was that it had been donated by Mr A. Wilson from Mount Wills in 1909.

Brazenor was excited by this discovery: Mount Wills in the Dandenong Ranges is some distance away from where the earlier possums had been collected, so it gave new hope for the chances of finding a living Leadbeater's Possum. The area around Bass River had been cleared for the most part and Koo-Wee-Rup Swamp had been drained, leaving little probable habitat for possums in any case. Brazenor was determined to track down Wilson and find out as much as he could about this mysterious possum. Having found no Wilson in any of the official records of the day, Brazenor travelled to the north-east of Victoria and laboriously called on every Wilson in the district, asking whether, thirty years ago, they might have sent a small possum to the museum. Wilson was never found, but the bush telegraph soon found Tom Hunter. In a letter to Brazenor, Hunter described how his gold-prospecting partner in Mount Wills had shot an unusual possum on the roof of their hut and sent it to the museum. Having pinpointed the actual origin of the Mount Wills possum, Brazenor scoured the thick forest-clad mountains near the old gold camp, but no Leadbeater's Possum could be found. It seemed that only a brief glimpse of this fascinating creature was permitted before it disappeared with barely a trace to join its relative *Burramys parvus* in extinction.

In 1960, a young geologist joined the museum as Assistant to the Curator of Fossils. Eric Wilkinson was an enthusiastic naturalist as well as geologist and a member of the Fauna Survey Group in the Field Naturalists Club of Victoria. In 1961, Wilkinson was surveying the fauna of the Cumberland valley, near Marysville – a popular tourist destination just over 100 kilometres north of Melbourne in the heart of the Mountain Ash forests. Forestry workers had reported seeing an unusual possum in this area, although it was far from where the earlier Leadbeater's Possum had been caught, and in the more heavily vegetated Mountain Ash forests. One night in April, Wilkinson and his colleagues saw what appeared to be a Sugar Glider; it scampered up into the foliage of a nearby Blackwood (*Acacia melanoxylon*) tree. With a clear view of the animal, Wilkinson noticed that it lacked the distinctive fluffy tail of the Sugar Glider and instead had a long thin tail with a brush at the tip. The only Sugar Glider-like animal to have a tail like this was the apparently extinct Leadbeater's Possum. Unfortunately the possum quickly scampered into the safety of the upper canopy and Wilkinson was unable to record any further details.

Nonetheless, Wilkinson was buoyed by the possibility of having rediscovered a living Leadbeater's Possum. He was busy planning further exploration of the area while driving home, when a nightjar flew in front of the car and perched next to the roadside. Ever the keen naturalist, Wilkinson stopped the car for a closer look when his spotlight caught a small pair of eyes flashing on a nearby tree trunk. The eyes belonged to yet another small possum, slightly browner in colour than the earlier one, but again without the fluffy tail of the Sugar Glider. This time, Wilkinson was able to confirm that it also lacked the Sugar Glider's characteristic gliding membrane. Wilkinson returned to the museum in high spirits, but notifying the director Brazenor that a lowly assistant had succeeded where Brazenor (now museum director) had failed, required some delicacy.

THE LEADBEATER LIFESTYLE

Leadbeater's Possum
(*gymnobelideus leadbeateri*)
G. Lewis

Ignorance about the behaviour and habitat requirements of the Leadbeater's Possum undoubtedly delayed its rediscovery. Searching for a small possum in the mountains and forests of eastern Victoria is no easy task, and it becomes almost impossible when nothing is known about the animal's habits. For example, when Charles Brazenor examined the five specimens in the museum's collections, he concluded that the Leadbeater's Possum was not 'built for speed' and would 'have none of the sprightly movements of the Sugar Squirrel [Glider]; it would move, rather, with the deliberate slowness and stateliness of the Ringtail [Possum] … Probably Leadbeater's is a scrub-dweller, spending most of its time in the underbrush and seldom ascending to the tops of the larger trees.' In fact, the Leadbeater's Possum is even more energetic and active than the Sugar Glider, and just as arboreal in dense forest.

Leadbeater's Possum has a restricted range at altitudes between 800 and 1300 metres in the Mountain Ash forests. The difficulty in finding it is due to its habit of living high in the canopy of the massive Mountain Ash trees. By watching dead trees at dusk when potential inhabitants emerge, current researchers can map the numbers living in the forest quite accurately.

Leadbeater's Possums conform to the typical possum pattern of living in trees and eating tree sap and insects. They also live in groups of up to twelve animals, as Sugar Gliders do. It might be expected, therefore, that these groups, as with the Sugar Gliders, would consist of a single adult male and one or more females and their offspring. So thought Andrew Smith when he began the first intensive study of a population of Leadbeater's Possums in 1980, but to his surprise he found the groups had a preponderance of males. The group is dominated by a breeding pair, who produce a distinctive musky odour and prolifically scentmark their territory of 1.5–3 hectares. The territory is primarily defended by the dominant female in the group; she aggressively excludes all other adult females, including her own daughters once they mature. Females require a territory in order to breed, and when one disappears from a territory her place is rapidly taken by an immigrant female.

Over the last two hundred years, the forest in much of the range of Leadbeater's Possum has deteriorated through age, fire and logging. Returning to the population studied by Smith, David

Lindenmayer looked for any associated changes to the possums' habits. Instead of the strict territorial exclusion of other females, Lindenmayer found up to three females nesting in the same box together. Nesting sites seem to have become so scarce in the forest that females have been forced to change their habits and share them. Smith also occasionally observed females in the same nest or territory for short periods, but noted that in each case only one successfully reared her young to independence. Whether or not more than one female is able to breed under these crowded conditions remains to be determined.

Wilkinson decided to delay making any announcement until further night surveys revealed more animals that he was able to photograph. With this evidence, Wilkinson showed his find to the delighted Brazenor who, as director, wanted an actual specimen before he made a public announcement. John Coventry, later Curator of Herpetology, was the museum's best marksman. Coventry and Wilkinson spent the next few nights departing for the forests straight after work, and returning, bleary-eyed, in time for work the next day. They were instructed to ring Brazenor, whatever time of night, as soon as they had obtained a specimen, which, with some relief, they soon did. With an irony inherent in many of the conservation functions of a museum, the Leadbeater's Possum was finally declared 'alive' on the incontrovertible evidence of a dead specimen.

The remarkable rediscovery of the Leadbeater's Possum could only be exceeded by the 1966 findings of some skiers who were staying at the University of Melbourne ski hut on the slopes of Mount Hotham in the Australian Alps. Ski lodge folklore claimed that the lodge had possums living in the oven. Sure enough, a small friendly possum emerged from the kitchen (if not the actual oven), but the skiers could not identify its species. The tiny mysterious bundle was carried down the mountain to the Fisheries and Wildlife Department, who could not identify it either. Eventually, an opinion was sought from Norman Wakefield (1918–1971), a well-known Victorian naturalist who had been working on some *Burramys* fossils found in Victoria. The mysterious possum was brought into the Museum of Victoria. Despite its good nature, the possum objected to having its lip raised in order to see if it shared the characteristic ridged molar of the *Burramys* fossils. After finding no refuge up the tall and lanky Wakefield and nearly escaping into the vaults of the museum basement, the possum was finally examined. Wakefield confirmed that the small animal was indeed a living *Burramys* and, for the next nine months, Fisheries and Wildlife Officers searched the mountains high and low for a mate for their captive possum.

THE MOUNTAIN-PYGMY POSSUM

Mountain Pygmy Possum

The unusual habits of *Burramys* (now known as Mountain Pygmy-possums) at first confounded searchers. No-one with a knowledge of Australian possums would have predicted that this tiny creature nests in the boulder fields on the upper slopes of the mountains. Most possums live in trees. *Burramys* not only prefers rocks, but lives in a region with few trees of any kind. Once the habitat requirements of *Burramys* were known, they were relatively easy to find. In the early 1980s concerns about alpine developments, and the effects these might have on the presumably tiny population of *Burramys*, led to an intensive survey of all suitable habitats in the alpine region. The entire known population of *Burramys* (probably no more than 2600 individuals) was found to occupy isolated pockets totalling just 10 square kilometres on the boulderfields of the Australian Alps in New South Wales and Victoria. Here, for six months of the year, these marsupial 'dormice' curl up in underground nests, emerging only when the snow recedes.

In the generally hot and dry Australian continent, *Burramys* and the echidnas (*Tachyglossus aculeatus*) are the only true hibernators. In many respects *Burramys* is similar to the northern hemisphere dormice and hibernating squirrels. Although *Burramys* takes every opportunity to feast on insects (particularly the fat and tasty Bogong Moth, which finds cool summer shelter under rocks in the high country), it also collects hard seeds and caches them. During hibernation, *Burramys* often make short excursions from their sleep-chambers to check the weather, find warmer sleeping quarters or enjoy a mid-winter snack.

Unlike their northern hemisphere counterparts, *Burramys* have strictly segregated sleeping quarters. Female *Burramys* live up to thirteen years in the wild – a long time for a small animal and considerably longer than the two to four years the males usually live. The larger females dominate the best feeding grounds high up the mountains and facing south, while males are forced to make do with the lower, north-facing slopes. It is a harsh world for any small mammal, let alone a male Mountain Pygmy-possum forced to live in poor habitats away from the best sources of food – and away from the females with whom he must mate. Perhaps because they are not able to stock up as well as the females, males emerge from hibernation more often. Males also come out of hibernation earlier than the females. It may again be hunger that drives this, but the lower, warmer boulder fields also thaw earlier than those occupied by the females, enabling the males to emerge early and make their way up the mountains to the female territories. Unlike females, males take time to attain breeding condition; while the females luxuriate in a long lie-in, the males have a chance to get in shape for the strenuous mating season ahead.

At Mount Higinbotham in Victoria, the main road to the ski village divided the female breeding area from the male hibernation area, making it impossible for the two populations to meet. Following strategies developed for badgers and frogs in Europe, tunnels were built under the road way. Artificial boulder fields connected the two ranges through the rock-filled tunnels. Since *Burramys* almost exclusively travel within their protective boulder fields, this effectively channelled the males through from their hibernation area to the female breeding grounds. The success of this so-called 'tunnel of love' was evidenced by the recording of a *Burramys* using the tunnel within an hour of its opening.

Since other pygmy possums lived in tree hollows, searchers assumed that the possum had been brought up to the ski lodge in some firewood. Searches therefore concentrated on forested areas below the ski lodge where firewood was collected. But no more possums were found.

It was not until the 1970s that more *Burramys* came to light. Three scientists from New South Wales had succeeded in trapping a *Burramys* in a rocky, dry streambed on Mount Kosciuszko in 1971. One of these scientists, John Calaby (1922–1998), observed that the foundations of the University Ski Lodge consisted of a coarse, rocky substrate. He suspected that the possums had been attracted to the ski lodge by the rocks, and not brought up to the area in firewood as previously thought. Calaby suggested that Joan Dixon, Curator of Mammals at the Museum of Victoria, concentrate her efforts to find *Burramys* in Victoria in similar rocky areas, rather than woodland. Dixon was soon successful in capturing a lone female, an event that won front-page coverage in the state's broadsheet newspaper. This new possum acquisition lived out its days in the pampered comfort of Dixon's lounge room, where its gentle temperament won the heart of its erstwhile curator. By 1972, enough *Burramys* had been captured to establish a breeding population in captivity – to which they adapted readily. By 1978 only thirty individuals had been captured in the wild.

Both Leadbeater's Possum and *Burramys* are remnants of larger populations that were once much more widely distributed when Australia's climate was different. Broom's original fossil discovery was in New South Wales. Fossils of Leadbeater's have also been found in other caves in New South Wales and Victoria. These fossils reveal that the Leadbeater's Possum evolved about 15–30 million years ago at a time when the wet sclerophyll forests of *Eucalyptus* and *Acacia* were expanding to take the place of contracting temperate rainforests. The fossil record seems to suggest that during cool, dry glacial periods when dry forests prevailed, Leadbeater's Possum was rare or absent; but during warmer, wetter interglacial periods, it was more abundant and widespread. The most recent glacial period (11,000–25,000 years ago) saw the wet sclerophyll forests shrink and fragment along the south-eastern seaboard of Australia. This change may have caused the Leadbeater's Possum to become extinct in much of its former range. Over the last 5000–10,000 years a warmer and wetter climate has seen the forests re-expand and Leadbeater's Possum may have become more abundant accordingly.

These possums both have a tenuous grip on life, with each species' total habitat smaller than most major cities (700 square kilometres for Leadbeater's Possum and a mere 10 square kilometres for *Burramys*). Pressure on these habitats has only increased in the last 200 years. In this relatively short space of time, Lindenmayer estimates that suitable forest for Leadbeater's Possum has shrunk from 44,000 square kilometres, covering the entire south-eastern tip of Australia, to a tiny forest stronghold. Leadbeater's Possums make their home in one of the most valuable timber forests in Australia. The Mountain Ash forests have an estimated value of over $200,000 per hectare and employ nearly 2000 people in the region. The circumstances of the Leadbeater's Possum closely parallel that of the Northern Spotted Owl in the forests of north-west America. Clearfelling not only destroys the possum's immediate habitat needs, but fails to provide any opportunity for the future development of suitable habitat. Until recently, old hollow trees were actively removed from forests because they were regarded as a risk to forestry workers and a fire hazard. Plantations of Mountain Ash are also unsuitable for Leadbeater's Possums, as they are usually harvested at about 50–80 years – long before they begin to develop hollows.

The shrinking range of the Leadbeater's Possum is not entirely due to forest clearance. Like many Australian plants and animals, the abundance of the species seems to follow the stage of burning and regrowth in the forest cycle. This possum depends on dead trees for hollows in which to nest. Mountain Ash forests typically burn every few hundred years or so,

Deforestation, particularly of Mountain Ash, has reduced the potential range of the Leadbeater's Possum to less than 2 per cent of its estimated pre-colonial habitat. S. Madder.

The majority of Australia's arboreal mammals and many of its birds are dependent on tree hollows for shelter and breeding. S. Madder.

and the mature trees killed in this process have abundant tree hollows for the Leadbeater's to utilise. The current Mountain Ash forest is the result of a major conflagration in 1939 which scorched 10 per cent of the state of Victoria, killing seventy people and clearing most of the Mountain Ash forests. But the old trees are now starting to decay and fall down. The regrowth trees are only in their sixties and will not start to develop hollows for another 150 years. Unlike in the past, today there are few contiguous forests at a different stage of the fire–regrowth cycle for the possums to move to.

Certainly, Leadbeater's Possum has not always been dependent on Mountain Ash forests – the Bass River region at the beginning of the twentieth century was a low, scrubby forest of Ribbon or Manna Gum (*Eucalyptus viminalis*), Swamp Paperbark (*Melaleuca ericifolia*) and tea-tree (*Leptospermum* spp.). Recently, small populations of Leadbeater's Possum have been found in similar lowland forests and also at higher altitudes in the Snow Gum forests. So long as suitable nest sites are available, there is a chance the possums may be able to shift to new forests, provided (and this is a big 'if') there are sufficient forests left.

The future for *Burramys* is in some senses more uncertain. Burramyids probably evolved about the same time as Leadbeater's Possum (15–20 million years ago). A number of *Burramys* species lived in the warm, temperate forests and later in the cool rainforests. As the climate cooled and dried, these forest species died out. *Burramys parvus*, the last of its kind, retreated upward from the forests to the sub-alpine woodlands and then to the alpine refuge to which it is now restricted. *Burramys* is an alpine and subalpine specialist, restricted to a narrow band of habitat 800 metres wide around the permanent tree line. In our current climate, the timber line lies at a mere 1800 metres which, given that Australia's tallest mountain is only 2228 metres high, provides just a handful of isolated peaks suitable for *Burramys*. Presumably, in colder times, *Burramys* has been able to extend its range much further, but genetic research by Megan Osborne at Museum Victoria suggests that this has not happened for a very long time. The three major populations of *Burramys* are so distinct genetically that they seem to have been isolated from one another for over half a million years. These populations remained on their peaks even though, at the height of the last ice age (a mere 25,000 years ago), the timber line was as low as 700 metres. Perhaps the fact that *Burramys* seem to spend their entire lives within the boulderfields of these high altitudes further restricted their potential range even when appropriate temperatures were found across the entire south-eastern Alps.

THE MAHOGANY GLIDER

Remarkable though the stories of *Burramys* and the Leadbeater's Possum are, they are not unusual in Australia. In the last forty years, a dozen or more mammal and bird species which were once thought extinct have been found alive in small, isolated remnants of their original habitat. All of them remain threatened by habitat clearance, introduced predators and competitors. The most recent rediscovery, the Mahogany Glider (*Petaurus gracilis*; Plate 16), was made after a labelled skin was uncovered in the vaults of the Queensland Museum in 1984. Unlike earlier specimens, this skin provided detailed locality data. Researchers hurried to the site near Cardwell in North Queensland where, just before the onset of the rains, they were fortunate enough to glimpse what were clearly Mahogany Gliders in the upper canopy. When they returned the following dry season, the farmer had cleared that last patch of forest for sugar cane. Mahogany Gliders have since been found in other small forest remnants, but certainly this is another species living close to the edge.

Tourist developments in the Alps, particularly skiing, directly threaten the remaining habitat of *Burramys*. Concerted lobbying and management plans have succeeded in protecting much of the threatened habitat and restoring some damaged areas, but winter sports are popular and profitable. Both skiers and *Burramys* must share the same small patch of snow in a generally hot and dry continent.

In the longer term, however, Australia inches ever closer to the equator. It seems increasingly likely that the world is experiencing a rise in temperature through global warming. Both natural and unnatural climatic change may conspire to thaw the last remaining patch of snow on which *Burramys*, and many other alpine species, depend. In the relatively flat Australian landscape there are no higher peaks for *Burramys* to retreat to. Australia has never been as hot or as dry as it is now, and for the foreseeable future that pattern looks set to continue. It may be only a matter of time before *Burramys* once again joins the ranks of extinct Australian mammals and we will have lost another link with a colder past.

Human activity undoubtedly threatens the continued existence of both Leadbeater's Possum and *Burramys*. But human activity did not make either of these possums rare in the first place. Like so many other species, both possums put their money on black when climate change in Australia spun to red. The very adaptations which make the *Burramys* such an exceptional alpine specialist, and the Leadbeater's Possum a dense-forest specialist, now imprison them in fragments of their original habitat. Adaptation is a double-edged sword. It benefits the species in the short term, but can prove fatal if the environment changes in the future.

Both species have survived such detrimental environmental changes in the past. Their populations are sustainable at current levels, but they are extremely vulnerable. If a conflagration like the 1939 fires again roars through the Mountain Ash forests it could wipe out the entire population of Leadbeater's Possum. A major landslide could destroy an entire population of *Burramys*. Disease, climate irregularities and habitat loss or degradation all represent greater threats to the reduced populations of these possums than they do to their more abundant cousins. The loss of a few thousand Sugar Gliders would be unfortunate, but the species would probably recover. The loss of a few thousand Leadbeater's Possums or *Burramys* would be the loss of the species. Human activities did not make these possums rare, but their rarity now makes them worryingly vulnerable to the environmental changes wrought by human activity.

PLATE 16. The Mahogany Glider (*Petaurus gracilis*) was thought to be extinct until rediscovered in Queensland. It remains highly endangered.
B. Cowell, Queensland Museum.

PLATE 17. The Weedy Seadragon (*Phyllopterax taeniolatus*) or Syngnathus was one of the more remarkable and beautiful marine creatures collected by Péron on Bruny Island. It was described in 1804 by Bernard Lacepède at the Muséum d'Histoire Naturelle in Paris from Péron's specimens.

By Ludwig Becker, from Frederick McCoy, *Prodromus of the Zoology of Victoria*, Decade VII, Plate 65.

Part 3

FROM FOSSILS
AND BONES:
THE LAST
250 MILLION YEARS

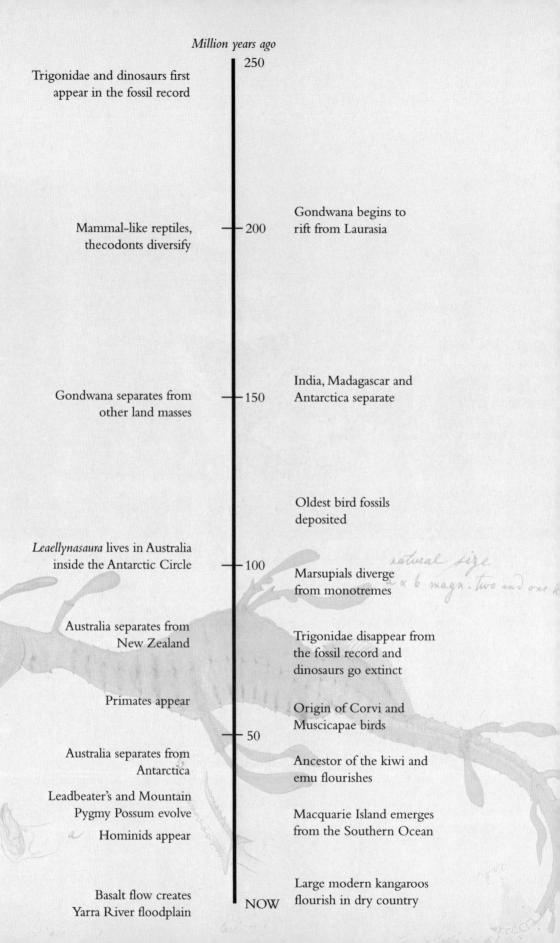

Million years ago

250 — Trigonidae and dinosaurs first appear in the fossil record

200 — Mammal-like reptiles, thecodonts diversify — Gondwana begins to rift from Laurasia

150 — Gondwana separates from other land masses — India, Madagascar and Antarctica separate

Oldest bird fossils deposited

100 — *Leaellynasaura* lives in Australia inside the Antarctic Circle — Marsupials diverge from monotremes

Australia separates from New Zealand — Trigonidae disappear from the fossil record and dinosaurs go extinct

Primates appear — Origin of Corvi and Muscicapae birds

50 — Australia separates from Antarctica — Ancestor of the kiwi and emu flourishes

Leadbeater's and Mountain Pygmy Possum evolve — Macquarie Island emerges from the Southern Ocean

Hominids appear

Basalt flow creates Yarra River floodplain — NOW — Large modern kangaroos flourish in dry country

7 THE CASE OF THE MISSING MOLLUSC

Neotrigonia margaritacea, the first living trigonia to be described.
Conchology Collection, Museum Victoria.

FROM THE SHORT perspective of human history, it seems obvious that we are in the midst of another great mass extinction. The planet has lost one vertebrate species each year (on average) for the last 300 years, with every indication that the toll will continue at this rate well into the future. This speed of extinction among vertebrates is extraordinary. Dinosaurs are estimated to have gone extinct at a rate of one single species every thousand years.

But the five mass extinction events that characterise Earth's fossil record exact their toll not just on vertebrate species but on invertebrates as well, particularly marine species. In contrast, human impact on invertebrates and marine species is relatively mild and the current extinction event has a long way to go before it can rival the 96 per cent of marine species that disappeared at the end of the Palaeozoic era.

What current events would even be apparent to the palaeontologist of the future? It is possible that none of the current bout of extinctions would even be noticeable given the patchy and erratic nature of fossilisation. But surely the massive proliferation of humans (from 300 million in 1 CE to 6100 million in 2000 and rising) would be noticeable? Surely the spread of humans around the world, in company with their host of companion species (grasses, rats, cattle, dogs, cockroaches) would be visible in the fossil record? Probably. But plenty of other species have proliferated, expanded and dominated the Earth in greater numbers and for far longer periods of time than humans – before disappearing, declining or retreating to a small corner of the world they once dominated.

The trigonia are one such family of creatures whose rise and fall would barely have been noticed were it not for the auspicious timing of their rediscovery, at key moments in the development of evolutionary theory. The trigonia offer a pertinent insight into the way in which the study of ancient lineages influences the way in which we interpret the natural world – and the way in which the irregularities in the fossil record can change our interpretation.

Trigoniidae were a diverse and striking family of molluscs common when dinosaurs roamed the Earth. In fact, the presence of certain trigoniids can be used to date the rock in which they are found. Trigoniidae are characteristic of the Mesozoic period. They first appeared in the fossil record 245 million years ago and were thought to have disappeared 65 million years ago during the major extinction that decimated the dinosaurs. These unusual shells were, for many years, regarded as relics of just another fascinating creature that had once lived on Earth but was long since extinct.

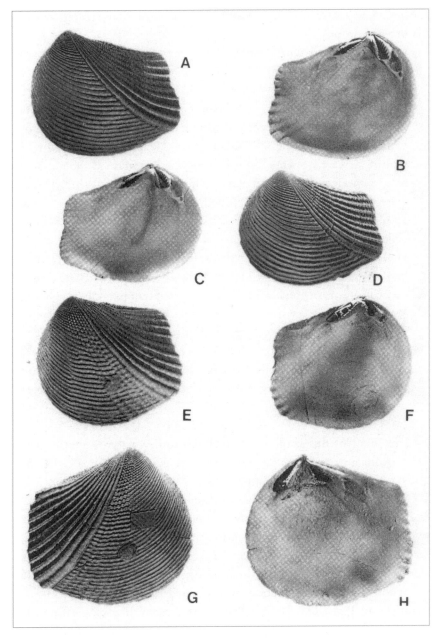

Like many cockles and clams of today, these fossil trigoniids had strong ridges on their shells. Unlike most modern molluscs, the ridges could run both ways in a pattern called 'discrepant sculpture' – half the shell might have concentric ridges while the other half has radial ridges. Many striking forms evolved, but even those without the strange bisected shells had a characteristic that readily distinguishes them from all other bivalves: distinctively flanged teeth in the hinge which holds the two halves together.

T. Darragh.

That was, until 1802 when the extremely well-equipped French scientific expedition under the command of Nicolas Baudin (1754–1803) arrived in Australia. During their two-year exploration of the southern Australian coast, Baudin and his naturalists collected more than 2500 species of animals. For the 'pupil zoologist' of the expedition, François Péron (1775–1810), Australia was a wealth of new and unknown species – every handful of shells he picked up on the beach included new

François Péron was 'pupil geologist' with Baudin's scientific expedition.

species, new forms and living representatives of forms long thought extinct in Europe.

One morning in what would later be named Tasmania, Péron went ashore for a day of collecting.

On the 20th at day-break, I embarked in a boat which was going to fish on the isle Bruny: I collected about twenty new kinds of fish, among which were two species of Lophies, two Ostracions, a Uranoscopos, a Cottus, a Raie or skate, two Scienae, the Antarctic Chimera, and a second species of the same genus, which was very remarkable from a bone on the top of the head shaped like a club; a Syngnathus adorned with several floating membranes, resembling so many streamers. I also collected a dozen or fifteen kinds of shells either entirely new or very rare, among which I found a valve of trigonia (*Trigonia Antarctica*, N.) a kind of shell never till now known to be a living subject; and which in our climates form such long banks of petrifications; and the beautiful Venus, with transverse ribs extremely fine and thin, brittle, and slight; as also divers Phasianelles of the greatest beauty, some elegant trochus, several turbots, one of which I have described under the name *Eustomiris*, reflected the most brilliant and lively prismatic colours; some kinds of patelles, fissurelles, and oscabrions, &c. &c. were the produce of this day's research.

The sight of this haul, which includes some of the southern coast's most exquisite creatures, was enough to make the terminally ill chief zoologist on board the expedition weep.

The trigonia was unmistakeably related to the fossil forms so abundant in Europe. But this shell was no fossil: it was the hard remains of a recently living animal. Its flanged hinge declared it unmistakably a trigoniid, but it was clearly a different species from any of the fossil bivalves with which Péron was familiar. The trigonia, along the with bulk of the other collections from the expedition, were lodged at the Muséum d'Histoire Naturelle in Paris. This large collection of Australian zoological material strongly influenced its curator, the prominent evolutionary thinker Jean Baptiste de Lamarck (1744–1829).

Lamarck named the species *Trigonia margaritacea*, after its lustrous interior ('margarita' means pearl) – a feature not apparent in fossils. The discovery of an extant species of Trigoniidae sparked intense scientific interest, particularly in France, and there was immediate enthusiasm for collecting a live specimen. In 1826, another French expedition came to Australia. Permission for *L'Astrolabe*, under Captain Dumont d'Urville

L'Astrolabe on the reef at Tongatapu, by Louis de Sainson from Dumont d'Urville's atlas (1830–35).

(1798–1842), to travel along the southern and eastern coasts was politely granted by the British Secretary for War and Colonies, despite the official state of war between England and France. However, Earl Bathurst also sent an urgent message to the New South Wales colony to send ships south pronto to establish a British presence in the area.

The French ship arrived in Westernport Bay, near present-day Melbourne, with much anticipation, and not, as the British suspected, because they were canvassing a possible colony site. *L'Astrolabe*'s charismatic naturalist and surgeon, Charles Gaimard (1793–1858), and his fellow naturalist/doctor on board, Jean-René Quoy (1790–1869), were looking forward to the possibility of finding a live trigoniid in Bass Strait. But they were to be disappointed. Dredging failed to find any live Trigoniidae. Only the separated shells of *T. margaritacea* gave tantalising testimony to the existence of living companions in the deeper entrance to Westernport. The journal of Quoy and Gaimard records that they 'left Westernport with the regret of not having found what we were looking for'.

L'Astrolabe sailed through Bass Strait and up the New South Wales coast without incident. The southward currents rushing down the east coast carve a deep channel just offshore. In depths of over 2 kilometres, there was no opportunity for further dredging until the crew of the *L'Astrolabe* found themselves close inshore under Cape Dromedary, 100

kilometres south of Sydney. The captain reported that 'the drag-net was thrown and pulled back several times; among the curious objects that M. Quoy found was a small, live trigonie, a shell for which he had been looking for some time'.

The remainder of the voyage of *L'Astrolabe* through the South Pacific yielded a productive harvest of scientific information on new and little-known species such as the Bare-backed Fruit Bat (*Dobsonia moluccensis*) and the Burrowing Bettong (*Bettongia lesueur*). None of these strange creatures, however, were as valued to Quoy and Gaimard as their tiny, unremarkable 'trigonie'.

We now know there are six living species of Trigoniidae – all of which are found only in Australia. They are all fairly small, cockle-like shells with strong, slightly striated ribs running down the outside and a lustrous 'mother-of-pearl' interior. Their purpley-green interior has even led to their use in jewellery. Visitors to beach-side towns in Victoria once took home souvenir salt spoons with shiny purple interiors and spiky backs. These, along with other souvenirs, were made from trigoniids, which the locals call 'brooch-shells'.

The first living trigoniid to be discovered, *Trigonia margaritacea*, is a thick, heavy shell, found in deep waters around the coasts of the south-east mainland and Tasmania. The very rare *T. strangei* was named in 1854. Perhaps in recognition of Lamarck's identification of the first extant trigoniid, the next living trigoniid species to be found (in 1858 near Sydney) was named *T. lamarckii*. Nine years later, the only living tropical trigoniid was identified, *T. uniphora*. In 1907 a Western Australian species was described, *T. bednalli*; like many marine organisms of the region, it exhibits small but regular changes in shape the farther west it is found. A second south-east coast species, *T. gemma*, joined the list as recently as 1924.

SHIPWRECKS AND SEASHELLS

During its travels, *L'Astrolabe* was grounded on a reef in Tonga. Dumont d'Urville was concerned that the ship might be broken up by rough seas before it could be refloated on the tide. He nailed down the ship's weapons in the bilge so that they would sink with the ship, feeling it safer to be unarmed than to deal with either islanders or crew in possession of weapons. All non-swimming members of the crew were instructed to board boats for shore with only 'their most necessary belongings'. The subsequent bundles of 'necessities' were so large, however, that they had to be reduced and repacked by the officers. In the midst of this mayhem, the naturalist Quoy sat calm and collected on the quarterdeck, writing up his notes as though nothing untoward was occurring. In his pocket, however, was the precious 'trigonie'. He later wrote, 'We were so anxious to bring back this shell with its animal that when we were, for three days, stranded on the reefs at Tonga-Tabu, it was the only object we took from our collection'. This trigoniid (also a *T. margaritacea*), along with the earlier shells, was sent back to Lamarck at the Muséum d'Histoire Naturelle in Paris.

Classification of the trigoniids, living and fossil, is confused by multiple names and classifications, as will become apparent later in the story. Tom Darragh from Museum Victoria recently revised the taxonomy of the Australian Trigoniidae. The trigoniids of the Mesozoic are all *Trigonia* species, while the living species belong to a different genus, the *Neotrigonia*. But in the early half of the nineteenth century, when the first living trigoniids appeared, all Trigoniidae were called *Trigonia* and debate raged over exactly how similar these living creatures were to their fossil predecessors.

The rediscovery of living trignoniidae coincided with a period of intense development in theories of evolution. Péron's trigonia had landed on Lamarck's desk at the very time when he was developing his evolutionary theory of transmutation (1803–1809). Lamarck was one of the most prominent pre-Darwinian evolutionists and he seized on the *Neotrigonia* to illustrate his views. In his 'transmutationist' view of evolution, individuals respond to changes in the environment and pass on successful adaptations to their offspring. One consequence of Lamarck's model is that it allows species to adapt to environmental change quickly, rather than

Péron's trigonia was received in Paris at the Muséum d'Histoire Naturelle by the curator for invertebrates, Jean-Baptiste Lamarck, famous for seeking to explain the origin of species through transmutation. This is Lamarck's drawing of the specimen.

taking the many, many generations required in Darwinian evolution. Extinction, therefore (in the absence of the dramatic environmental changes wrought by humans), should not occur. The reappearance of the Trigoniidae justified Lamarck's claim that 'Small species, especially those that dwell in the depths of the sea, have the means to escape man; truly among them we do not find any that are truly extinct.'

Darwin's theory of natural selection was published fifty years later, in 1859. Natural selection fundamentally differed from transmutation as a mechanism for evolution. While transmutation relied on the inheritance of learned or acquired traits (and hence allowed for rapid change of organisms in response to environmental change), the theory of natural selection admits only the inheritance of congenital traits. Natural selection is thus a vastly slower process. Extinctions under the Darwinian model should occur through slow and gradual competition between species, with the more successful species gradually increasing in abundance and the less successful declining to small pockets of populations in isolated outreaches of their original habitat.

The Trigoniidae seemed to offer an excellent example of this process. At the height of their abundance, Trigoniidae species numbered in their hundreds and were found throughout the world. Darwin found them restricted to a handful of species in the farthest Antipodean outreaches of the British Empire (and where could be more isolated than that?). 'A single species of Trigonia, a great genus of shells in the secondary formation, survives in the Australian seas; the utter extinction of a group is generally, as we have seen, a slower process than its production.'

But what happened to the trigonia between their disappearance from the fossil record 65 million years ago and their sudden reappearance as living species? No fossils had been found at all during the Cainozoic period (the so-called age of the mammals). This 'missing link' was a major problem for evolutionary proponents, who argued that every modern species was linked in a continuous chain of ancestry back to a common ancestor. Evolutionary theory does not allow for species to go extinct and then reappear (albeit in a modified form) centuries later. Both Lamarck and Darwin simply assumed this absence was due to the paucity of the fossil record.

In comparison, creationists seized on the Cainozoic gap as a fatal flaw in evolutionary theory. Louis Agassiz (1807–1873) was one of the most ardent and authoritative creationists. In 1840 he argued:

> The absence of *Trigonia* in Tertiary [Cainozoic] strata is a very important fact for discussions on the origins and relationships of species of different

epochs; for if it could one day be shown that Trigonia never existed throughout the entire duration of Tertiary time, it would no longer be possible to maintain the principle that species of a genus living in successive geological epochs are derived from each other.

And quite correct that is: it would disprove evolutionary theory completely except that, geology being what it is, it would never be possible to prove that Trigoniidae never existed at that time. The fact that fossils do not exist does not mean that a species did not exist, but if a Cainozoic fossil Trigoniidae was ever found, Agassiz would be proved wrong. Not that he would have accepted that verdict. In an eloquent piece of bet-hedging, Agassiz argued that even if a continuous lineage of fossils could be found, it would merely demonstrate that God may have his own reasons for allowing continuity between groups of similar species across time.

Eventually the missing Cainozoic Trigoniidae was found. A geologist, Richard Daintree (1832–1878), was conducting some routine mapping for the Victorian Geological Survey in the coastal cliffs of Torquay, south-west of Melbourne, when he uncovered some fossils. The Geological Survey sent all fossil material to the National Museum of Victoria in order to date the rocks they were mapping. Similar fossils were uncovered a short time later by the Survey in cliffs at Beaumaris, now a Melbourne suburb.

Frederick McCoy reviewed the material for identification. McCoy recognised these fossil Trigoniidae as a new species and realised from the

Charles Darwin (1809–1882). His theory of natural selection arose out of extensive observations of the natural world, ranging from the mundane (earthworm activities) to the exotic (coral reef formation) to the obscure (the trigonia).

Bird Rock Cliffs, west of Jan Juc beach, Torquay, taken by Richard Daintree while surveying the area in 1861 for the Victorian Geological Survey. State Library of Victoria.

The trigonia was displayed at both the Melbourne Exhibition of 1861 and the London Exhibition of 1862. Illustrated here is an earlier Melbourne Exhibition, 1854.

State Library of Victoria.

surrounding fossil material that they were not from the Mesozoic period of other trigoniids, but from the Cainozoic. He named the fossils *Trigonia semiundulata* and placed them on display in the 1861 Melbourne Exhibition. He even mentioned them briefly in his essay accompanying the exhibition. But McCoy wrote no formal description of the new species. He provided no published account of how to distinguish this species from other previously described species and hence *T. semiundulata* remained 'unofficial' in the accepted international process of naming and describing new species.

McCoy saw the Cainozoic trigonia not as a link in the evolutionary chain, but as a vindication of his own particular view of Australian natural history. Many scientists of the time argued that Australia was some kind of refuge for ancient fauna. But for living trigoniids to be some kind of relict of Mesozoic species implied a continuous chain of descent through the ages with ongoing modifications resulting in quite different living species of *Neotrigonia* – in other words, evolution. As a creationist, McCoy would have none of this. He pointed out that, despite their similarities, the Cainozoic trigonia is a different species both from those found earlier in the fossil record and from modern species. McCoy believed this was evidence that the Cainozoic trigonia was 'created' at this time, and was neither a descendant of earlier species nor the ancestor of living trigonia. McCoy strongly rejected the idea of Australia as a refuge for Mesozoic creatures.

Part 3 FROM FOSSILS AND BONES

WAS IT A COVER-UP?

McCoy's failure to publish the Cainozoic trigoniid discovery – such an important piece of evidence for evolutionary theories – might be interpreted as a deliberate attempt to ignore facts contrary to his creationist views. Stephen J. Gould observed that McCoy 'must have known what he had and what it meant. But he didn't even bother to publish his description'. McCoy was an ardent creationist like many prominent Australian biologists.

In 1861 McCoy would certainly have known of (and opposed) Lamarck's transmutationist view of evolution. (It was fully developed by 1809 and widely disseminated in English through the critique contained in Charles Lyell's seminal *Principles of Geology*, published in 1830.) But Darwin's *On the Origin of Species* arrived in Sydney only in 1860. Darwin was not in contact with any Australian scientists until the 1870s and so his views were not widely discussed prior to the book's arrival. Although McCoy was certainly opposed to evolutionary theories, he was probably not familiar with natural selection in 1861. In 1870, McCoy expressed his views on the 'order and plan of Creation', repudiating 'progressive development' and 'Darwin's theory of successive evolution of different species by "natural selection"'.

McCoy certainly recognised the importance of his discovery for opponents of evolution. In 1875 he noted that, prior to his discovery of the Cainozoic trigoniids, the family had 'furnished an extraordinary apparent exception to the usual distribution of genera in time, according to which a genus living in the older periods of the world's history, and becoming extinct during a subsequent geological period, is not found to reappear at a still more recent epoch'. But if his intention was to hide the Cainozoic trigonia, he would hardly have put it on display in the 1861 Melbourne Exhibition nor sent it to the 1862 London Exhibition.

McCoy's failure to publish his findings on the trigonia probably stemmed from his preoccupation with his *Natural History of Victoria* (later renamed *The Prodromus of the Zoology of Victoria*. McCoy rarely published small papers and, as a consequence of his position at the museum, he received a large volume of new specimens to describe and publish, many of which did not appear until many years later.

T. semiundulata was seen at the London exhibition by H. M. Jenkins, a little-known British geologist (of a Darwinian persuasion). He could not conceal his glee at so obvious a vindication of Darwin's argument that the Cainozoic gap was merely an artefact of the imperfect or imperfectly investigated fossil record. Jenkins published a small paper drawing attention to this remarkable discovery, and thus he inadvertently became the 'official' discoverer and namer of the species, even though he mistranscribed McCoy's term *T. semiundulata* as *T. subundulata*, the name by which this species is now known.

Jenkins' interest in this discovery increased when he obtained some fossil material from Cainozoic rocks in Australia. Jenkins acknowledged the argument that *T. subundulata* was so different from both the fossils of antiquity and the modern forms that they may not represent a clear line of descent. Indeed, Darragh's review of recent work agrees that *subundulata* is unlikely to be the ancestor of modern *Neotrigonia*. But Jenkins was surprised to discover fossil trigoniids which he regarded as indistinguishable from the living species *T. lamarckii*. Here was more compelling evidence of a direct line of descent for the Trigoniidae captured in the fossil record.

McCoy may have been mildly irritated by Jenkins' mistranscription and poor description of his Cainozoic trigonia, but the suggestion that there was a fossil version of a living species caused him to react with uncharacteristic alacrity. A mere six months after Jenkins' article appeared, McCoy felt 'urged to make a preliminary publication … on account of Mr. Jenkins' paper' explaining that this fossil was in fact a new species, *T. acuticostata*, and not the same as the living species at all.

The incomplete nature of the fossil record does indeed make evolution difficult to demonstrate directly. Where there is a similarity between two species in the fossil record, can we assume that one is descended from the other with modification over time? Only a series of intermediary stages, suggesting a gradual modification of one form to another, can illustrate a clear line of descent. Although trigoniids were once found on many continents, only Australia has a record of Trigoniidae spanning from the Mesozoic forms to the living species. Even so, the fossil record is at best patchy, with certain key chapters missing altogether.

Darragh's review of the taxonomy of the Australian Trigoniidae suggests that the modern forms, *Neotrigonia*, are probably descended from one of the older, more widespread *Eotrigonia*, which in turn is probably descended from one of the trigonia. Thomas Hall (1858–1915), from the University of Melbourne, noted that Trigoniidae share a very similar juvenile form, no matter how different they may look as adults, and the

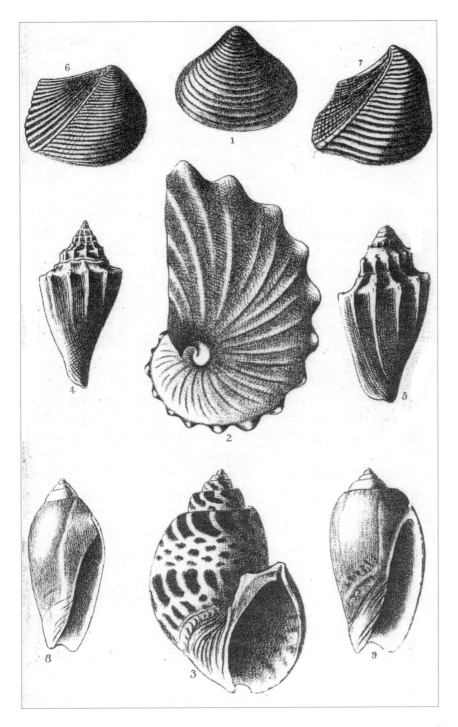

Illustration of *Trigonia subundulata*.
H. M. Jenkins (1865).

juvenile *Neotrigonia* look very much like *Eotrigonia*. The juvenile form of one fossil, *Eotrigonia eocenica*, recently discovered by Darragh, bears a strong resemblance to some living *Neotrigonia*. *E. tubulifera* also closely resembles

Neotrigonia and, if it is not the actual ancestor of living species, it might be a very close relative of an ancestor. But there is no clear intermediary form which is as similar to *Neotrigonia* as it is to the *Eotrigonia*.

If McCoy reviewed the evidence today, he would probably see no reason to change his view that different periods of the Earth's geological history contained their own similar, but distinct, species of Trigoniidae. McCoy might, however, have much greater difficulty explaining the multitude of other families which also show continuity of form (to a greater and lesser extent depending upon the completeness of the fossil record) from one geological age to another. McCoy's view of how life on Earth had arisen shaped the way in which he interpreted everything he saw in his professional work. Such fundamental theories are not to be altered on the basis of one small whimsical mollusc family. McCoy might continue to ask why, if trigoniids are descended from one another, are the trigoniids across the ages quite different species? But the other, perhaps more pertinent, side of this query is, why are the different species of Trigoniidae across the ages so similar, if they are unrelated?

The Trigoniidae had the good fortune to be rediscovered in the midst of the development of great evolutionary theories. With impeccable timing, the first living species appeared just as Lamarck was formulating his transmutationist theory. The hotly debated missing link trigonia then appeared almost as a vindication of Darwin's newly published natural selection theory. But the trigonia's story is as much a warning about the foibles of the fossil record as a vindication of evolution. Great claims are often made on the basis of a few fragments of fossilised shell or bone. Such scanty evidence is sometimes inconclusive one way or the other and can be interpreted to suit a variety of alternative views. In science, as in life, nothing is certain but uncertainty, and the most attractive theory can be swept aside by the slow and gradual accretion of opposing observations.

PLATE 18. *Quantasaurus intrepidus* is another small dinosaur from south-eastern Australia that lived in the cold, often frozen conditions 110 million years ago. Its bones show no signs of growth rings, suggesting it may have ben warm-blooded rather than hibernating through the winter.

P. Trussler.

PLATE 19. Whether they are 'four-handed', 'finger-footed' or have two hands and a pair of feet, all apes (including humans) share more similarities with each other than they do with monkeys. G. Cuvier (1817) *Le Regne Animal.*

8 BRAINBOX

P 185990

Fossil imprint of the brain of the dinosaur
Leaellynasaura amicagraphica, collected at Dinosaur Cove,
Victoria, by Tom Rich and Pat Vickers-Rich.
Vertebrate Palaeontology Collection, Museum Victoria

A HUGE, LUMBERING sauropod slowly raises its tiny head. It peers myopically into the distance before ripping the top off a nearby palm tree and munching contentedly in a bovine-like trance. Big, slow and dumb – such is the common image of the dinosaurs.

Admittedly, modern reconstructions on film and television like *Jurassic Park* and *Walking with Dinosaurs* have broadened the popular image of dinosaurs. Big and small dinosaurs have been depicted in a much broader range of behaviours than the traditional munching herbivores or crunching predators. Flocking *Gallimimus* ostrich-dinosaurs demonstrated social cohesion, calling sauropods advertised their desire to communicate, and the intelligent velociraptors provided a fearsome example of matriarchal hunting packs in *Jurassic Park*, for example. Although physiological reconstructions are often detailed and sophisticated, behavioural reconstructions offer greater challenges. We may know what they looked like and how they moved but do we know how these creatures really lived and behaved? The simple answer is that we can only guess. Using tiny clues they left behind, drawing parallels with living creatures and filling in the gaps with a lot of educated guesswork, palaeontologists can hazard a reasonable supposition. But this is by no means a clear-cut or simple process. Reconstructing dinosaur life from even the most detailed fossil requires a lot of interpretation.

REDUCE, REUSE, RECYCLE

Evolution is a frugal process. It recasts old features for new functions. It reuses, remoulds, and reshapes. Dramatic adaptive mutations are rare, and in the meantime evolution makes do with what it has. As a consequence, dramatic changes in behaviour are not always reflected by changes in physiology. For example, the general physiology of a Ferret (*Mustela putorius*) and a Mink (*M. vison*) (pictured left) is almost identical. They have the same kind of fur, the same type of feet, the same eyesight and the same breathing patterns. And yet while Ferrets generally avoid water, the Mink is so aquatic it spends its entire life in or around rivers, streams, lakes or seas. There is nothing in a Mink's physical features that would lead anyone to think it was aquatic. Certainly no palaeontologist would be able to deduce the Mink's lifestyle from its fossil skeleton alone. The same basic body plan has been adapted for a dramatically different lifestyle.

The semi-aquatic American Mink (*Mustela vison*).

Dinosaur remains are relatively scarce in Australia, with much of what we know being derived from footprints rather than fossils. Although the large sauropods did roam the Australian landscape, they left little behind. For some reason, the most common fossil dinosaurs found are those of small dinosaurs, such as the kangaroo-sized hypsilophodontids (pronounced *hip-sil-OFF-a-don-tids*). Some were as small as 2–3 kilograms – about the size of a small turkey or the smallest wallabies. The southern Australian hypsilophodontids were among the smallest dinosaurs known.

These dinosaurs were not only relatively small, but they were probably also fast. Hypsilophodontids, like kangaroos, had large hindquarters, small forelimbs and a long tail. The structure of these large legs and the muscle attachments show that they were undoubtedly used for fast locomotion. And why should all dinosaurs have been slow when their nearest living relatives, the birds, include ostriches which can cover 60 kilometres in an hour? Even large cold-blooded reptiles like crocodiles can outrun a human over 100 metres. No living animal with a hypsilophodontid kind of body plan is slow – big, powerful, bipedal legs are always for moving fast.

THE REPTILE WITHIN

Fish and reptiles generally rely on a fairly simple brain divided into sections broadly associated with particular functions (such as olfactory lobes for scent, optic tectum for vision, cerebellum for regulating movement). In contrast, most mammals (and particularly humans) have developed a 'new brain' – the neocortex – over the top of this more primitive brain. The neocortex also co-ordinates many of the functions controlled by the old brain in non-mammalian vertebrates. But the old brain is not completely superseded. It still performs components of the functions for which it once had sole responsibility. For example, when the visual cortex is damaged, some people report that they cannot see anything at all. And yet if they take a guess at where a light might be, they point directly to it. This subconscious 'blindsight' is enabled by the optic tectum, which once co-ordinated visual activity in the reptilian brain. The presence of a particular structure (like a human appendix, for example) does not mean it necessarily performs a particular function – it may be a harmless left-over. Such left-overs make the task of the palaeontologist even more difficult.

Having dispensed with the 'big, slow' part of the dinosaur myth, what can we deduce from fossils about dinosaur intelligence? Nerves do not usually fossilise, and soft tissue disappears, leaving no trace. Well, almost no trace. Encased in its own bony hardhat, the brain often leaves a telltale impression on the inner surface of the skull. In fossil mammals and birds, the inside of the skull provides a detailed imprint (or endocast) of all the surface brain structures, allowing us to deduce what may or may not have gone on inside the brain when it was alive. One small Australian hypsilophodontid has left just such a record – *Leaellynasaura amicagraphica* (pronounced *lee-ellen-AH-saura AM-icka-GRAFF-icka*).

The kind of brain an animal had can tell us a little about how it interpreted and interacted with the world around it. And if one species has a brain that differs dramatically from those of its relatives, it suggests that a powerful and unusual environmental influence demanded a dramatic shift in physiology. Just such a dramatic shift is evidence in the fossil skull fragment of *Leaellynasaura amicagraphica*. The species was named after Leaellyn Rich, the daughter of palaeontologists Pat Vickers-Rich and Tom Rich. They found and studied the fossil on the Victorian coast of south-eastern Australia at Dinosaur Cove, among some of the richest dinosaur fossil streams in Australia.

The first distinctive feature of *Leaellynasaura*'s brain imprint is that it is very clear. Unlike most bird and mammal brains, modern reptile brains are often smaller than their skulls – generally filling only half of the available space. The impression left on the inner surface of the skull is indistinct compared to that of mammals and only vaguely reflects the shape and proportions of the brain. Dinosaur skulls generally show the same pattern, suggesting that they had brains similar in relative size to modern reptiles. But some members of the reptile family had detailed endocasts, suggesting that their brains filled the entire cranial cavity. The flying pterosaurs also had large brains for their skull size, similar in some ways to that of modern birds, as did a few of the meat-eating carnosaurs.

Leaellynasaura belongs to a moderately large-brained group. Hypsilophodontids were part of a larger group of herbivorous dinosaurs, the ornithopoda. Just as giraffes are related to other ruminants like deer, cattle, and sheep, *Leaellynasaura* was related to other medium-sized herbivores like the duck-billed hadrosaurs and the large *Iguanodon*. There are many descriptions of the brains of hadrosaurs and *Iguanodon*, and they do tend to have larger brains (in proportion to their body size) than the sauropods, for example. But none of the ornithopod endocasts have as clear an imprint as *Leaellynasaura*, suggesting that it had either an unusually large brain or an unusually small skull for this group.

THIRD EYE OR THE SEAT OF THE SOUL?

The pineal–parietal complex (sometimes called the pineal or parietal gland, body or eye) is one of the strangest parts of any vertebrate's brain, varying enormously in structure and function in different species. Its origins are not clear, but certainly it arose from an early visual sensory system. Some even argue that the parietal and pineal were bilateral light sensors or two primitive eyes on the top of the head. In some frogs and lizards, the parietal pokes up through a hole in the skull and has cells similar to the retinal cells of the eyes. This is the so-called third eye or parietal eye. The parietal eye probably helps cold-blooded animals to keep themselves warm (by basking or seeking shelter), and is reduced in tropical lizard species compared to temperate species. Because of the regular daily and seasonal cycles of temperature, the parietal eye is particularly sensitive to daily and seasonal rhythms.

Early in his career, Walter Baldwin Spencer investigated the evolution of the third eye in lizards such as the Lace Monitor. He noticed that it was most complex in conservative (or 'primitive') species like members of the monitor family, including the Australian perentie or goanna (*Varanus giganteus*) and Lace Monitor (*V. varius*). It was decreased in complexity in more recent species like the Asian dragon (*Calotes* spp.) and the even more modern chameleon (*Chameleo vulgaris*). Spencer did not consider any living species to have the highly developed third eye of some extinct species (labyrinthodonts, plesiosaurs, ichthyosaurs and iguanodons). He concluded that the third eye may be 'most rightly considered as peculiarly a sense organ of the pre-Tertiary periods'.

More recently evolved families lost their parietal eye as the hole in the skull closed up, leaving only an imprint of the pineal–parietal body on the endocast. But, despite its apparent separation from the outside world, the pineal–parietal complex remains as a seasonal time-keeper in many warm- and cold-blooded animals. In many mammals the pineal gland regulates behaviour such as hibernation, seasonal reproduction and migration, and also governs the secretion of hormones. The pineal–parietal complex also seems to be involved (along with the hypothalamus) in temperature regulation in mammals. Animals

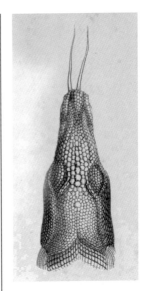

Detail of the Lace Monitor's head, showing the third eye.
Prodromus of the Zoology of Victoria, plate 41.

that have lost or dramatically reduced their pineal–parietal complex (like pangolins, armadillos, anteaters, sloths, echidnas and dugongs) have unusually low core body temperatures for mammals.

In humans, the pineal is implicated in both jetlag and seasonal affective disorder. We primates are the only species where the pineal–parietal complex is not visible on the endocast: our neocortex has completely covered the 'old brain', leaving the pineal gland almost in the centre of the brain. The seventeenth-century philosopher René Descartes proposed that the pineal body was the organ in which the human soul resided and that it regulated the activities of the rest of the brain (illustrated below). Never again has the pineal been lifted to such an exalted position in our understanding of how brains work.

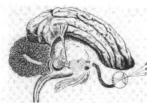

Descartes' erroneous location of the pineal gland in the hollow ventricles of the brain.

Part 3 FROM FOSSILS AND BONES

The individual *Leaellynasaura* whose brain endocast we have was not an adult. Judging from the larger femurs of other individuals which were found nearby, this particular animal was probably only about two-thirds adult size. Many creatures change shape and size dramatically during their life, and *Leaellynasaura*'s unusual brain might be a feature of her immaturity rather than a species characteristic. But only a few animals dramatically alter their brain structure after birth. Humans are highly unusual in being born with a brain that triples in size after birth. Lampreys spend their first five years blind, filter-feeding through river-mud, before growing teeth, eyes and an enlarged optic tectum in order to hunt fish at sea before breeding. Such transformations, however, are the exception rather than the rule and allow the brain to expand to its full potential in later life. For the young *Leaellynasaura* to develop a brain like that of other ornithopoda in adulthood, her brain would have to shrink. Such a developmental pattern is highly unlikely.

Endocasts only provide an impression of the surface structure of the brain, and the skull remnant of *Leaellynasaura* is only of the top of the skull. There are two features of *Leaellynasaura*'s brain that are strikingly different from that of her fellow ornithopods. *Leaellynasaura* shows a clear imprint of a pineal body, a feature that is usually absent in other ornithopoda. *Leaellynasaura* also had a remarkably large optic tectum for a dinosaur. In fact, the optic tectum of other ornithopods is so small that it does not even leave an impression on the skull surface. If we can understand what these structures do in living animals, their presence may explain a little of how *Leaellynasaura* lived and why she needed such an unusual brain.

Presumably dinosaurs did not have souls, for the pineal–parietal complex was so greatly reduced in dinosaurs that it is rarely apparent on the internal surface of the skull. The absence of this complex in dinosaurs has been taken as evidence that they could not have been warm-blooded and were probably ill-equipped to deal with extremes of temperature. Does *Leaellynasaura*'s pineal body (as distinct from the parietal eye of her reptilian cousins) signify a different strategy for regulating heat, like being warm-blooded, or hibernating, or migrating?

These questions can be answered only by examining additional information from her skeleton and surroundings. Is it just a coincidence that a dinosaur with such an unusual brain should be found in such an unusual environment for a dinosaur?

LEAELLYNASAURA'S WORLD

Dinosaur Cove is now part of the spectacular Victorian coast, famous for its eroded limestone pinnacles. But *Leaellynasaura* was excavated from the older mud- and sandstones of a very different world. One hundred million years ago, Dinosaur Cove was an inland streambed, probably washing bones and debris away from a large, freshwater lake (Plate 18).

Lakes like this one may have been vegetated by quillworts (*Isoetales*), hepatics and algal species. Patches of moorland surrounding the water catchment contained the frond-like straps of *Lycopodium* and sphagnalean mosses. The lakes and rivers teemed with aquatic invertebrates like mayfly, mosquito and sandfly larvae (Ephemiptera and Diptera). Crustaceans such as the Spiky Seed Shrimps (ostracods), Fairy Shrimps (anostracans) and Waterfleas (cladocerans) were abundant. Freshwater bryozoans and bivalves would have provided grist for the grinding jaws of the ceratodontid lungfish. Slow-moving, heavily scaled coccolepidid palaeonisciforms swam alongside more modern streamlined teleost fish (koonwarrids and leptolepids), no doubt both the prey of long-necked freshwater plesiosaurs (more commonly found in marine ecosystems). Amphibians and turtles also occurred in the region.

The lakeside was surrounded by thick conifer forests much like the Hoop Pine forests of modern Tasmania. Ginkgoes (maidenhair trees) were also common. Ferns, mossy bryophytes and delicate fern-like sphenopsids covered the ground, while the drier forest edges were dominated by hardier sclerophyllous ferns. Species resembling modern cycads (*Taeniopteris daintreei*) and pandanus (*Pentoxylaleans*) formed the mid-storey. These cool, moist forests were home to a diverse assemblage of invertebrate life, including spiders, earthworms, beetles (coleopterans), flies and mosquitos (dipterans), bugs, cicadas and aphids (hemipterans).

A few fossil feathers indicate that birds were also present in these forests, although their precise identity remains a mystery. Remains of the gliding pterosaurs are more common. Heavily armoured ankylosaurs and perhaps even a horned neoceratopsian (related to *Triceratops*) may also have roamed the region. Unlike other localities of a similar age, this area is dominated by the hypsilophodontids, perhaps as many as four or five different species. Carnosaurs no doubt terrorised the small hypsilophodontids. The fleet-footed and elegant ornithomimosaurs or ostrich-dinosaurs may have been less concerned, relying on their impressive turn of speed to out pace predators.

Australia lay much further south than its current location, inside the Antarctic Circle attached to Antarctica. In fact, the south pole, with its periodic meanderings, may even have been on Australia, rather than Antarctica, during this period. Oxygen isotope data from rocks surrounding *Leaellynasaura* suggests average annual temperatures of 0–8°C. The evergreen trees may have protected themselves from cold with thick bark, while deciduous plants preserved energy during cold, dormant seasons by losing their leaves. Fossil lycopod plants show patterns of growth suggesting annual periods of dormancy. Modern Isoetes now live in the alpine bogs of Tasmania where temperatures regularly drop below freezing in winter. But many ferns and bryophytes have no such resilience to cold and their abundance suggests there must have been some pockets within the forest ecosystem where the temperatures did not drop below 10°C.

This evidence suggests that the flood-plain on which *Leaellynasaura* lived was cold, even freezing, in winter. More significantly the region, being just inside the Antarctic Circle, was probably shrouded in total darkness for at least one month of winter.

If, as research suggests, *Leaellynasaura's* environment was seasonally cold and dark, how did she survive? Perhaps she migrated north for the winter? It is possible that polar dinosaurs from Alaska may have migrated south. Migration requires regulation of behaviour by seasonal rhythms (a key function of the pineal body) as well as requiring good spatial orientation and mapping (perhaps through the optic tectum). *Leaellynasaura* was, however, only a small animal and small animals do not typically migrate long distances unless they can fly. In order for *Leaellynasaura* to reach climates that were markedly warmer than where her fossils were found, she would have to travel thousands of kilometres each year. The direct path north was blocked by an inland sea in the Great Artesian Basin, requiring a detour west before travelling north. The Alaskan dinosaurs (which, incidentally, lack a pineal body) were as large as modern migratory ungulates, like bison and deer, and had an unobstructed north–south route available to them.

Leaellynasaura's small size makes her an unusual polar animal. Most species that live close to the poles are larger (with smaller surface areas relative to their mass) than their relatives in warmer climates. *Leaellynasaura* was a small dinosaur, even by hypsilophodontid standards. Cold-blooded animals are also rare in polar regions. But there is a surprising diversity of strategies employed by modern cold-blooded creatures to cope with cold. The New Zealand Tuatara (which even in the age of dinosaurs was a 'living fossil' 100 million years old with a well-developed parietal eye) is the most cold-tolerant of living reptiles and remains active at temperatures as low as 5 degrees C. Some North American frogs, like the Wood Frog (*Rana sylvatica*) and Spring Peeper (*Hyla crucifer*) hibernate in rock cracks and crevices that freeze up in winter. The high concentration of glucose (just like anti-freeze) in their vital organs allows these frogs to stop breathing and their hearts to stop beating and yet they thaw out in spring without ill-effect.

While *Leaellynasaura's* environment may have been a borderline temperature for a cold-blooded reptile of her size, it would not be a particularly harsh environment for a warm-blooded animal of a similar size. Was she warm-blooded? The 'hot-blooded dinosaur' debate has raged long and hard, with the weight of opinion finally resting with the argument that most dinosaurs were probably cold-blooded. Most dinosaurs, however, were very large, whereas *Leaellynasaura* was quite

High levels of glycopeptides allow the ghostly Antarctic Icefish (*Champocephalus gunnari*) to live in sub-zero seawater while its large heart and thin blood carry oxygen that cannot be carried by red blood cells at such low temperatures.

Part 3 FROM FOSSILS AND BONES

small. In regulating body temperature, size is pivotal and small reptiles have more in common with small mammals than they do with large (especially gigantic) reptiles when it comes to keeping warm. Small size is, in fact, regarded as a prerequisite to becoming warm-blooded, despite the fact that some warm-blooded animals have subsequently become quite large. Many members of cold-blooded families have developed varying levels of temperature regulation. Many fast-swimming fish (like mackerel, sharks and tuna) maintain higher blood temperatures than their surrounding environment. Some turtles maintain high body temperatures by using their large body size as a 'heat sink'. Being 'warm-blooded' does not have to involve the highly regulated temperature control typical of humans with their finicky temperature-sensitive cortex. For creatures with smaller brains and less cortical material, internal temperature is less critical; many mammals survive happily within a much broader range of body temperatures than humans can.

The pineal–parietal complex plays a pivotal role in thermoregulation for most animals. Animals without a pineal–parietal complex or with very small structures (like the Echidna, *Tachyglossus aculeatus*) find it difficult to deal with extremes of temperature. In cold conditions, echidnas go into torpor. Other small mammals faced with cold may hibernate to reduce their energy use during lean winter months. Hibernation allows animals to lower their basic energy requirements to the minimum necessary for survival. Their body temperature drops and all unnecessary body functions cease. Such 'non-essential' functions include even the laying down of bone. Most hibernating animals show distinctive patterns of bone deposition which coincide with annual cycles of activity. But *Leaellynasaura*'s bones show no such rings – she seems to have been active all year round.

Without the option of hibernation, *Leaellynasaura* may have been forced to expand the size of her pineal–parietal complex to improve her ability to control her body temperature in a cold climate. No matter how primitive *Leaellynasaura*'s system for keeping her core body temperature above that of her environment, it would have been an enormous advantage in a cold climate. Warm-bloodedness – or, at least, unusually good temperature control – may have allowed *Leaellynasaura* to exploit a polar habitat that excluded her cold-sensitive predators (like the crocodilians) and competitors (like the sauropods).

But what of *Leaellynasaura*'s second unusual brain feature – that enlarged optic tectum? What was it about her environment, or her response to it, that required the development of such an unusual expansion?

THE BRAIN'S MAPMAKER

The optic tectum (also called optic lobes or superior colliculus) is primarily responsible for vision in reptiles. In conjunction with *Leaellynasaura*'s large eyes (typical of hypsilophodontids), her large optic tectum signifies a creature strongly dependent upon visual stimuli. But the optic tectum also interprets information from other senses, including sound, smell, infra-red light and electro-magnetism as well as balance and the body's own sense of location.

The optic tectum has layers representing the different senses. These are arranged like map overlays that provide different information about the same place. Sound and visual information from a particular external location activate neurones in the same vertical column in the optic tectum, enabling a co-ordinated response to the information. When you glimpse a ball out of the corner of your eye, turn towards it and stretch your hand out to catch it, your optic tectum helped to co-ordinate the whole process. When you hear squealing brakes, turn to see a speeding bus and leap back onto the footpath, again it is your optic tectum that has processed the sights, sounds and smells that tell you where that moving object is and how to avoid it. The optic tectum is the brain's navigation centre, interpreting and coalescing complex spatial maps of information from all the senses, enabling you to orient towards, and avoid or approach, moving objects.

The optic tectum is clearly vital when important components of an animal's visual world are moving. Frogs use their optic tectum to catch flies and flee from predators. Flying birds typically have a large optic tectum to help them negotiate their rapid movements through three dimensions. Similarly, arboreal mammals typically have a larger optic tectum than their ground-dwelling relatives. Among reptiles, the flying pterosaurs had a large optic tectum.

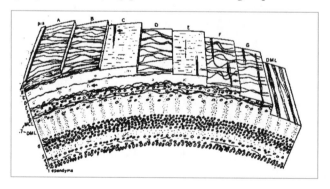

Cells of the optic tectum and optic fibres in the frog.
H. D. Potter

The co-occurrence of a dinosaur that apparently lived at least part of its life in darkness with an enlarged optic tectum suggested to Tom Rich and Pat Vickers-Rich that *Leaellynasaura* may have been nocturnal. Low light requires improved ability to detect movement and discriminate contrasts, and this may have required a larger optic tectum.

However, the opposite argument could also hold. In low light there is less visual information to process and, although there is certainly selection for large eyes (to maximise light input), there is less information to process and hence less need for a large optic tectum. Nocturnality in primates and birds is associated with increased size of olfactory parts of the brain (rather than vision). In contrast, diurnal primates (which are active during the day) devote more of their brain to processing visual information than their nocturnal counterparts. Nocturnal reptiles also typically show a heavy reliance on other senses, such as heat-sensors, rather than vision. Diurnal reptiles have the largest optic tecta, while burrowing species with reduced eyes have the smallest.

It seems likely that the same general patterns hold for both mammals and reptiles. Diurnal species have highly developed eyes and visual processing centres in their brains. Nocturnal species that still utilise vision have larger eyes and pupils but may begin to reduce the size of the visual layers of the optic tectum. Once light levels decline to a point where vision becomes ineffective, eye size and tectal size decrease dramatically, particularly in burrowing species where visually modulated spatial maps also become unimportant. *Leaellynasaura*'s large optic tectum provides no definitive evidence for the idea that she may have been nocturnal.

What other features of *Leaellynasaura*'s environment might require enhanced visual acuity or spatial mapping of movement? Either *Leaellynasaura* needed to move rapidly through a complex three-dimensional world, or she needed to keep track of something else which was moving rapidly.

Leaellynasaura did not fly – that much is clear from her skeleton – but ever since T. H. Huxley first described hypsilophodontids from the Isle of Wight, there have been suggestions that they might be arboreal or tree-dwelling. Arboreal mammals certainly have a larger optic tectum than their terrestrial counterparts. However, when Peter Galton reviewed the evidence he found nothing in their skeletons to suggest that hypsilophodontids were any more likely to be arboreal than other dinosaurs. He did, however, accept that any small and active dinosaur could well have climbed trees. The lack of specific adaptations for living in trees does not rule out that possibility.

Tree-kangaroos (*Dendrolagus* spp.), for example, are highly arboreal and yet have similar ground-dwelling adaptations to many of their terrestrial cousins (particularly rock-wallabies). In fact, despite the ease with which

Reconstruction of a 1869 hypsilophodontid from the Isle of Wight.
T. H. Huxley (1869).

they negotiate the treetops, tree-kangaroos look remarkably incongruous with their large feet and stiff, muscular tail and often upright posture. One of the skeletal features of tree-kangaroos that does distinguish them from ground-dwelling kangaroos is a large cavity in the forearm joint allowing greater mobility of the forelimbs when climbing trees. Hypsilophodontids also have a large forearm space, although not larger than any of their clearly terrestrial relatives. Such a feature in hypsilophodontids may not have been 'caused' (or selected for) by living in trees, but could well provide a pre-adaptation for living in trees.

The final option is that *Leaellynasaura* needed her large optic tectum to track and react to fast-flying objects. As she was only the size of a small turkey, she could have been prey for large flying predators like the pterosaurs. But this seems unlikely as the dominant predators at the time appear to have been the small terrestrial theropods. It is more likely that *Leaellynasaura* ate fast-moving flying prey such as insects. The larvae of flying insects were abundant in the lakes and rivers where *Leaellynasaura* was found. Insects are also seasonally abundant in many sub-polar regions today.

Hypsilophodontids are typically regarded as herbivores because they have basic cutting and grinding teeth. Teeth, however, can be misleading. Modern mammals often have such highly specialised teeth that this feature alone is used to classify different species into particular groups. Although such specialised teeth clearly reflect diet, 'generalist' teeth may not. Humans, for example, like other primates, have generalist teeth, which probably evolved for an omnivorous diet of fruit and vegetables. Yet humans have a very high level of meat in their diet. Birch mice have typically rodent teeth for eating grass and seeds, but they feed almost exclusively on insects. Eating insects actually does not require particularly specialised teeth. Most reptiles eat insects whole; one or two bites to crush the exoskeleton may be the only processing required. *Leaellynasaura's* typical hypsilophodontid teeth would have equipped her to take advantage of a seasonal, or even year-round, supply of high-energy insects to sustain her high metabolism in cold weather.

Unless more extensive material comes to light (and that is unlikely given the scarcity of dinosaur fossils in Australia) we may never know how *Leaellynasaura* really lived. Even the most careful and logical reconstruction based on skeletal, environmental and neuronal information can be wrong. But there must have been some reason for her remarkable brain, which cannot be explained either in terms of her hypsilophodontid heritage or her immaturity. It seems that *Leaellynasaura* was a specialist who stood out from the herd. In the dawn of the Cretaceous, a small, quick dinosaur may well have darted through the cold, dark Australian forests, keeping warm on a feast of abundant insects.

9 THE
APE CASE

The Ape Case at Museum Victoria were the first gorillas to be seen
in Australia and excited great public interest and debate.
Illustrated Melbourne Post, 25 July 1865, State Library of Victoria

HUMANS ARE NATURAL egotists. We tend to think of ourselves as the best and the brightest of all living things, whether creatures in God's image or creatures at the pinnacle of evolution. It is hardly surprising that the public furore which erupted in the nineteenth century over Charles Darwin's theory of natural selection was largely concerned, not with the implications for fossil dinosaurs or obscure molluscs, but with the displacement of humans from their self-appointed pedestal (Plate 19).

The ever-cautious Darwin almost completely refrained from discussing humans in the context of evolution and natural selection in his initial work, *On the Origin of Species*, in 1859. But he was well aware of the implications of his theory for human evolution. The theory of natural selection was the product of long years of accumulated observations of the natural world. To Darwin, there was no better alternative explanation of the facts – no matter what the implications for his religion or his species. Nothing would have pleased Darwin more than to have been able to plead a special case for humans, but his rational mind would not allow it. All logic suggests that humans are animals too, and the same universal laws of origin that apply to them must apply to us. Thirteen years after he published *On the Origin of Species*, Darwin gave in and published *The Descent of Man* in 1871, directly addressing the contentious issue of human evolution.

Although Darwin may have been wary about the consequences of extending his work into the human realm, his colleagues and supporters were not. Alfred Wallace (1823–1913) was prepared to extend their theory of natural selection to its logical conclusion – that natural selection can not only modify species, but actually create new species. as the progenitor of new species, even though Darwin preferred merely to comment on its ability to modify and alter species. Similarly, Thomas Henry Huxley (1825–1895), often referred to as 'Darwin's bulldog', was delighted to take Darwin's theory into the sensitive heartland of religious concerns. His arguments are expressed in numerous forms, perhaps most notably in *Evidence as to Man's Place in Nature* (1863), the title of which pre-empts its conclusion: that humans, like other animals, are a product of nature rather than divine intervention.

Although he was undoubtedly a fine scientist in his own right, Huxley became famous as a populariser of Darwin's ideas. But popularisation is fraught with dangers. There is a fine line between synthesising scientific concepts into a readily comprehensible argument and oversimplifying or even misrepresenting them. Huxley danced this line to his peril on the issue of the evolutionary 'progression' of humans. Also known as the ascent of man (literally, since such arguments often

Thomas Henry Huxley.
R. Barrett (1874).

Part 3 FROM FOSSILS AND BONES

APE ANATOMY

Ever since the great seventeenth-century anatomist Edward Tyson (1650–1708) completed his thorough investigation of the anatomy of a Common Chimpanzee (*Pan troglodytes*, erroneously labelled an Orang-outang or Pygmie) it has been clear that there are strong resemblances between the apes and humans (see Tyson's illustration of a Chimpanzee heart, right). As with his work on the Virginia Opossum, Tyson was ahead of his time with his careful, quantitative approach to knowledge.

Detail of the heart of a Chimpanzee (*Pan troglodytes*). E. Tyson (1699).

Tyson documented forty-eight features in which the chimpanzee is anatomically more similar to humans than to other apes and monkeys, and thirty-four features in which the chimpanzee resembles apes and monkeys more than it does humans. He concludes that 'our *Pygmie* does so much resemble a *Man* in many of its parts, more than any of the *Ape-kind*, or any other *Animal* in the World that I know of' (Tyson, 1699). So similar to humans was this creature that Tyson felt compelled to refute any suggestion that the chimpanzee might be a hybrid between humans and apes, although he did feel that the existence of such animals was the probably the source of ancient myths about human-like creatures.

Tyson's work greatly influenced scientists for the next two centuries and formed a keystone to the 'two great laws' of nature: unity of plan, and adaptation to particular environments. To any observer of natural history it was apparent that the animal kingdom consisted of variations on a common plan, and it was as an explanation of this puzzling phenomenon that Darwin offered his theory of natural selection.

place the sexes on an evolutionary, moral and intellectual progression with females on the lower side), this concept is a seductive one pandering to our need for special treatment. 'Evolutionary progression', and the erroneous use of living species to illustrate it, is a common and tempting way of simplifying evolutionary discussions, but it can lead to some blinding misconceptions.

Huxley's contribution to the debate about how humans fitted into the animal kingdom centred on our relationship to the apes. Clearly, if humans are closely related to any animal it is to the great apes: chimpanzees, gorillas and the Orangutan.

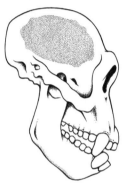

Chimpanzee skeleton
and skull.
S. Madder.

Huxley sought an answer to the question 'is Man so different from any of these Apes that he must form an order by himself? Or does he differ less from them than they differ from one another, and hence must take his place in the same order with them?' Clearly Huxley felt that humans belonged in the same group as apes, but towards the end of his discussion he begins to use the dreaded 'progressionist' shorthand that has confounded debates on the topic ever since. It was this shorthand which allowed the Anglican Archbishop Samuel Wilberforce to ask if Huxley was descended from an ape on his grandmother's as well as grandfather's side. Huxley's famous retort, that he would 'rather have a miserable ape for a grandfather [than] ... a man highly endowed by nature and possessed of great means of influence and yet who employs those faculties and that influence for the mere purpose of introducing ridicule into a grave scientific discussion' may have been clever, but it did nothing to clarify the real issue.

In discussing similarities and differences between humans and other primates, Huxley begins to place them in a series. His diagrams illustrate human teeth at the top of the page, followed by comparison with a gorilla, a baboon, a capuchin monkey and a marmoset. Skulls illustrate the same progression from lemur to baboon to gorilla to human. The pelvis progresses from the narrow one of the gibbon to the wider one of the gorilla to the broad-flanged human pelvis. And the human hand is shown in progression from the gorilla and orangutan. But perhaps Huxley's most famous imagery is the frontispiece of *Man's Place in Nature*, where the gibbon and stooped apes transform into the graceful upright form of the modern human. This image has been reproduced countless times, and it is often interpreted as implying that humans have evolved from the living apes.

Nor was Huxley shy of using this shorthand in his writing. The prime example of the relationship between the primates, Huxley argued, was in their brains. Although other scientists like Richard Owen (1804–1892) hotly disputed it, primate brains do share important similarities that are not found in any other group of animals. All primates have unusually large and complex brains for their size, and this expansion is particularly obvious in the neocortex (a feature which is absent in birds and reptiles but present in mammals). The expansion of the neocortex gives the characteristic wrinkling of the outer surface which makes the human brain look rather walnut-like. All primates share a certain pattern to these wrinkles (or sulci), just as all primate neocortices can be divided into three lobes – the occipital, temporal and frontal lobes. But rather than concentrate simply upon these similarities, Huxley was into full progressionist swing.

PROGRESS: A PART OF NATURE

Huxley's preoccupation with 'progression' was typical of his times. The concept of progress pervaded social and intellectual ideas throughout the nineteenth century. Auguste Comte described human history in terms of improvement from a primitive state to a perfect scientific society. Karl Marx developed a social theory based on the concept of human perfectability. In Adam Smith's economic writings, human society is depicted on a path of improvement. Herbert Spencer argued that 'progress…is not an accident but a necessity…It is a part of nature.'

Darwin's ideas about natural selection were seen as the biological manifestation of 'progress'. The survival of the fittest seemed to imply an increasing perfectability. But to what? In the nineteenth century, people were starting to come to terms with an ever-changing, unstable world, having emerged from the seemingly permanent and intransigent feudal and theological state. Darwin's notion of ever-changing species was comprehensible (just) within the constraints of a stable world, but even Darwin was not really ready to accept the full, fluid, unpredictable consequences of natural selection within an ever-changing physical environment. 'Progress' would be a consequence of natural selection if the environment for which species are being selected was constant – but it is not. Indeed, natural selection argues that progress in nature is entirely accidental, even though it is also a necessity. With an ever-changing environment, selection may operate first in one direction and then in another. There is no 'ultimate' end-goal, and hence there can be no concept of long-term progress.

The famous illustration of man and apes from T. H. Huxley's *Evidence as to Man's Place in Nature* (1863).

Perhaps no order of mammals presents us with so extraordinary a series of gradations as this – leading us insensibly from the crown and summit of the animal creation down to creatures from which there is but a step, as it seems to the lowest, smallest, and least intelligent of the placental Mammalia. It is as if nature herself had foreseen the arrogance of man, and with Roman severity had provided that his intellect, by its very triumphs, should call into prominence the slaves, admonishing the conqueror that he is but dust.

Thus, Huxley seems to argue that living mammals provide a clear example of evolution in the development of their brains, from the simple, primitive brain of tree-shrew, which is little more than a glorified smelling organ, to the advanced complexity of the human brain with its enormous neocortex. Huxley's passionate insistence on the strength of the similarities between primate brains leads him to overstate the case – from similarities between species sharing a common ancestor, to progression from one species to another.

In fact, there is really no such thing as a 'primitive' living animal. Every species is as highly adapted to its own particular conditions as any other. If life on Earth has evolved from a single source, at roughly the same time, then the lineage of any living species is as long as all the others. Those we typically term 'primitive' are simply species that seem to have retained certain features which appear very early on in the fossil record. For some reason, those features have not needed to change. In some senses, 'primitive' species can be regarded as nature's success stories – species that developed a winning formula early in their history which has sustained them for millions of years. Huxley may have regarded the tree shrew as having a primitive brain and humans as having an advanced brain, but it is more accurate to say that the tree-shrew's brain shares more features in common with its ancestor than the human brain does.

For simplicity's sake, we tend to use living species as examples of ancestral forms, and this gives rise to the fallacious lineages of living species so prevalent in Huxley's writings and diagrams. In truth, Huxley's diagrams simply lack the underlying family tree which would clarify the real relationship between the illustrated species. With this family tree, it is quite clear that humans did not descend from a gorilla, nor a gorilla from a baboon, etc. Rather, humans are descended from a creature that had certain features which both humans and gorillas and chimpanzees have inherited.

ANCESTRAL RELATIONS

How do we know certain brains are more like their ancestors than others? This simple illustration shows relationships between the brains of a solenodon insectivore (*Solenodon*), a hedgehog (*Erinaceus*), a pygmy tree-shrew (*Tupia minor*), a dwarf lemur (*Cheirogaleus*), a Sifaka lemur (*Propithecus*), a pygmy marmoset (*Cebuella*), a woolly monkey (*Lagothrix*), a gelada baboon (*Theropithecus*), a gorilla (*Gorilla*) and a human (*Homo sapiens*). Next to the brains are the evolutionary relationships of the species concerned, deduced from their fossil remains.

If we wish to know which features evolved first, we must look at all the living species and see which features they all share. All of these mammal brains share an olfactory bulb, a palaeocortex and a neocortex. Since it is unlikely (although not impossible) that

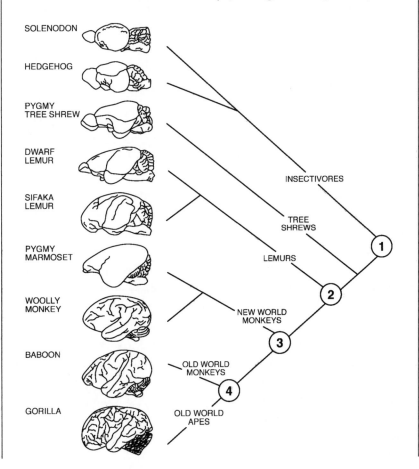

they all evolved the same features independently, the simplest explanation is that their common ancestor (at point 1) had a brain with all these features. Using the same process of deduction, we might imagine that ancestor 2 had a larger brain with a larger neocortex, while ancestor 3 had a neocortex almost completely enveloping the old brain.

Since our deductions about ancestral brains are often derived from living species (with limited input from the fossil record), it is not surprising that we slip into shorthand when discussing primitive and advanced brain states. Because insectivores share brain characteristics with the oldest ancestral state, we tend to describe them as 'primitive'. And because the human brain has the least in common with any ancestral form, we tend to describe it as 'advanced'. But 'advanced' implies some kind of ideal, a goal or ultimate purpose, and evolution has no goal except to allow animals that best suit a given environment to survive for as long as that environment exists. In an ever-changing world there is no advance towards a biological ideal nor away from a primitive state. The insectivore can more accurately be said to have 'retained' features common to its ancestor, while the human brain is a 'derived' feature – different from its ancestors.

George Britton Halford (1824–1910), head of Australia's first medical school at the University of Melbourne, Professor of Anatomy, Physiology and Pathology.
Melbourne University Archives

The brain is an example where humans have acquired a recent and unusual adaptation, while other primates have retained brain features more similar to our common ancestor. But the human brain is unusual – few physical features have undergone such dramatic change over such a short period of time. In many other features humans are arguably more 'primitive' than their primate cousins. For example, fossils suggest that the ancestral primates were more similar to apes than they are to modern monkeys. Human teeth resemble the teeth of ancestral fossils more closely than monkey's teeth do. Overall, the modern gorillas, monkeys and lemurs are just as different from their common ancestor as humans are.

But the shorthand progressionist fallacy is pervasive; at the height of evolutionary debates in the nineteenth century, it diverted considerable attention from the real issue. Instead of asking 'Are we related to the apes?' as Huxley originally framed his investigation, creationists debated the question 'Are we descended from the living apes?' Huxley's arguments aroused international opposition. One vehement opponent was George Britton Halford, a medical professor at Melbourne University, and the subject of his arguments was the hands and feet of apes and humans.

QUADRUMANOUS APES

Along with their large brain, bipedalism (or walking on two feet leaving the forelimbs free for manipulation) is a highly developed characteristic of humans. Indeed the evolution of these two features is probably not unrelated – the ability to manipulate, gesture, make tools and so on has often been argued to form an evolutionary feedback loop with brain size, the advancement of one feature driving the expansion of the other and vice versa. In the vast majority of mammals, all four limbs are used for locomotion. Even among animals that have forelimbs with highly developed manipulatory skills, few have completely desisted from using the forelimbs for locomotion – kangaroos, a few hopping rodents and humans are perhaps the only examples. In strict terms then, the terminal appendages to most mammals' limbs can be termed 'feet' and their characteristics are largely constrained by the need to bear weight and assist with locomotion. Forelimbs that terminate in manipulatory appendages can be regarded as having 'hands'.

A chimpanzee illustrating its manipulatory skills.

These definitions seem straightforward enough, but primates provide a clear example of the difficulties involved in determining what is a hand and what is a foot. Many primates (like the monkeys) use climbing as their primary mode of locomotion. Clinging to the round branches of trees imposes different constraints to those of running along the ground. Tree-dwelling favours similar foot adaptations to those favoured by the demands of manipulation; thus some monkeys may be argued to have four hands, rather than four feet. As a consequence, even the great apes like the ground-dwelling gorilla and chimpanzees were termed 'quadrumanous' or four-handed by some anatomists. The distinction between hands and feet is, of course, of particular fascination to humans, in whom hands and feet are so different. Early debates therefore often focused on the extent to which human hands differ from their counterparts on the forelimbs of other apes.

Huxley disputed the interpretation that all the apes were quadrumanous, or four-handed, while humans were not. He pointed out that the similarities between the feet of humans, gorillas and orangutans were far greater than the similarities between these species and the feet of the smaller, tree-dwelling monkeys. On the basis of this, and other structural similarities, Huxley concluded that humans did indeed belong in the same order as the great apes.

Halford also disagreed with the interpretation that the apes were four-footed creatures, favouring instead the term *cheiropodous*, or finger-footed. But the conclusion Halford drew from his anatomical analysis led him, not to a discussion of whether or not humans are related to apes, but whether or not humans could be descended from apes. 'Surely the intricacies of the Monkey's foot were planned as was also the comparative simplicity of Man's! They could never run the one into the other, or to use a fashionably scientific term be "developed" the one from the other.'

Despite Huxley's sloppy presentation of his position in the course of *Man's Place in Nature*, his conclusions do, in the end, revert to the relevant hypotheses that he formulated at the beginning of his book. Halford, however, had been diverted from the real debate into a fallacious one about the descent of humans from apes. In part, this error is compounded by Halford's own lack of precision. Halford's anatomical evidence came from a macaque (*Macaca*), a genus that contains some fifteen species and which is

The ape case display commissioned by Frederick McCoy, Museum Victoria.

Part 3 FROM FOSSILS AND BONES

a monkey, not an ape; it was apes with which most anatomists drew their comparisons. Halford, however, regarded his specimen as representative of the whole primate group, interchanging the terms *ape, simian* and *monkey* throughout his work. Strictly speaking, an ape is a member of the Hominoidae superfamily, while a monkey belongs to a different superfamily, both of which belong to the primate order and both of which are 'simian' (as distinct from the Prosimians, or lemurs and lorises). Not only are Halford's terms imprecise, covering 146 species, but they are actually mutually exclusive (an ape cannot be a monkey). Even at the time, when the taxonomic relationships were less clear and terminology less precise, Halford was criticised for his inaccurate turn of phrase.

Nonetheless, Halford's position as a prominent member of the Melbourne scientific establishment guaranteed his work an audience, and his disputation with Huxley was taken up by supporters and detractors in the local press, at public meetings and even in the National Museum of Victoria. Frederick McCoy supported Halford's anti-Darwinian stance by providing specimens with which to illustrate his public lectures. Despite the museum's policy of not loaning material for such purposes, McCoy felt that illustrating new points for lectures was 'the best use of such specimens' and instructed his taxidermist to deliver the material to Halford in time for his lecture and return it afterwards to public display. In 1865, McCoy added a powerful visual image to Halford's arguments with the acquisition of a group of stuffed gorillas (opposite), the first to be seen in the colony. Promoting this new display, McCoy exhorted visitors to see

how infinitely remote the creature is from humanity, and how monstrously writers have exaggerated the points of resemblance when endeavouring to show that Man is only one phase of the gradual transmutation of animals; which they assume may be brought about by external influences and which they rashly assert is proved by the intermediary character of the gorilla between the other quadrumana and man.

Attendance at the museum over the following week more than doubled. If this statement were not enough to make McCoy's allegiances abundantly clear, he also advised visitors to pay particular attention to the hand-like nature of the gorilla's feet – an obvious reference to Halford's work.

The scientific precision of both Huxley and Halford were put to shame a few years later by the thorough and detailed work of Huxley's former student, anatomist St George Mivart (1827–1900). There is no ambiguity in Mivart's hypothesis, which he tested by examining the

The hand of an
anthropoid.

anatomical similarities (skeletal and soft tissue) between humans and every known species of primate. 'Whatever existing species is most nearly related to that extinct root-form, which, according to Mr. Darwin's hypothesis, was the immediate ancestor of man – must exhibit a greater number of structural characters like those of man than any other existing species.'

Mivart noted that much of the dispute about hands and feet depended upon definitions, and he preferred the anatomically distinct terms of *manus* and *pes* to refer to the terminal appendages of the fore and hind limbs respectively. 'Hand' and 'foot' carry functional definitions about levels of dexterity or locomotion which confuse debates over their structural similarity. Although Mivart felt that no-one should be criticised for calling apes 'quadrumana' he did believe it was misleading, since if all the apes and monkeys were classified as quadrumanous, so too must humans. The differences between the hands and feet of humans and apes were distinctly fewer than the differences between the apes and monkeys. Mivart emphatically concluded 'that of Apes and Lemuroids (as well as of man) it must be said that each and all they are severally provided with "TWO HANDS AND A PAIR OF FEET"'.

Mivart also supported Huxley's contention that humans belong in the same group as the other large apes (now known as the family Pongidae, comprising gorillas, the Orangutan, chimpanzees and gibbons). However, Mivart was less convinced by the popular notion that gorillas in particular were more closely related to humans than any of the other apes. In this, Mivart felt that his research failed to show any evidence in support of evolution or natural selection, and his writing begins to show signs of the progressionist fallacy.

Mivart noted that, although humans show many similarities with the great apes which they do not share with the lesser apes (gibbons), they also show similarities with some of the lesser apes which they do not share with greater apes. In this he agreed with McCoy, who had earlier noted that in particular instances 'the lower kinds of monkeys … show in reality much nearer approaches to the human types of structure' than the great apes, although no-one would consider monkeys to be more like humans than apes. Determining which similarities indicated common descent and which were clearly independent developments of similar features seemed to Mivart to be impossible. And indeed it is a difficult task and one which occupies a great many taxonomists to this very day, even with the assistance of complex computer programs. For example, is the protruding human nose evidence of our close relationship with the big-nosed Proboscis Monkey (*Nasalis lavartus*)? To Mivart the body of data revealed no overwhelming evidence in support of a gradual transition from monkey to ape to human. In this he was correct, but many would today dispute the latter part of his conclusion that primate relationships present 'a tangled web, the meshes of which no naturalist has as yet unravelled by the aid of natural selection'.

Perhaps Mivart was simply overwhelmed by his data which would require the comparison of innumerable temporary family trees, each based on a single suite of characters. Although noses might place humans and proboscis monkeys in the same group, brain structure would place humans with the great apes, for example. These various family trees (which could number hundreds, even with Mivart's data) need to be compared and coalesced into the tree which offers the simplest explanation for the most features, while those features which fail to fit into the common family tree (like the human nose) are placed to one side as possible instances of independent or convergent evolution. Without this kind of analysis, anatomical data does indeed present rather a messy web of real and apparent inter-relationships.

But Mivart was also influenced by the fact that he was looking at second-hand evidence. The shared characteristics he sought were the theoretical result of a common ancestor many millennia in the past. In the meantime, each species had evolved independently, lost common features and gained unique ones of its own. And some of those acquired features are the 'independent' characteristics that presented so much error to Mivart's data. If we could compare a known ancestor (which can only be found in fossil form) to a known descendant, much of that error would be removed and a vastly clearer pattern of transition should emerge. Unfortunately, that option is not usually available, and the fossil evidence of human evolution is particularly patchy. Seeking evidence of a transition in living species is far more difficult. It is rather like looking for evidence of a familial resemblance in third cousins when your hypothesis is that parents pass on some of their looks to their children.

To Mivart, natural selection offered no useful explanation for the independent acquisition of features, or convergent evolution. But in fact it does. Where creatures live in similar environments, with similar environmental pressures, they may well evolve similar responses to that pressure. Thus, both the mammalian dolphin and piscean shark share similar body features, not because their common ancestor was also the same shape, but because they are both fast-swimming water-dwellers. The expected evolutionary transition of one form to another is confounded by the very principle of natural selection which Darwin proposed to explain how such a transition might occur. Halford himself stumbled across this environmental confounder of straightforward inheritance, when he

Australopithecus africanus was the first hominid to show clear evidence of consistent bipedal locomotion. It is the scanty fossil evidence that provides the strongest insight into human evolution, rather than comparative studies of our fellow apes.
K. Nolan.

described the monkey's finger-foot as being 'perfect, exquisitely beautiful in its anatomy, it is yet for walking an ungainly, useless member; for clambering, climbing, running from bough to bough, sleeping on the unsteady tops of trees, it is supremely fitted and ever to be relied upon'.

Halford's words were apt. Natural selection predicts that any species will be well-adapted to its particular environment but that its ability to adapt will be constrained by the particular body plan it has inherited from its ancestors. Humans may have evolved an extraordinarily large brain, but despite its large neocortex the human brain is still recognisably a primate one. Gorillas have a vast belly allowing them to digest large quantities of low-quality fodder, but they still have a recognisably primate digestive tract. Other apes are not some relict of human evolution or even a glimpse of our past (any more than humans are a glimpse of the past of other apes!); they are like distant cousins, living in a far-off country with an entirely different language and culture. There is certainly a family resemblance, but the differences are just as interesting and tell us every bit as much about how evolution operates.

VISIONS OF
NEW WORLDS:
THE LAST 4.5 BILLION YEARS

Million years ago

5000

Creation of rock in Murchison
and ALH84001 meteorites

Formation of the sun

Planets formed

— 4000

Erosion channels
formed on Mars

Colonial microfossils
in Western Australia

Liquid pours over rock on Mars
depositing carbonate globules
First evidence of photo-synthesis
in South Africa

First occurrence of
stromatolites

— 3000

First pockets of oxygen
appear on Earth

Earth's interior becomes active
and plate tectonics begin

End of banded ironstone and
beginning of sedimentary red-beds

Primitive bacteria at Bitter
Springs, Northern Territory

— 2000   Oxygen builds up to
appreciable levels

Ice sheets cover the Earth for
400 million years

—1000

Cambrian explosion of
life on Earth

Meteor wipes out
85 per cent of life on Earth

NOW   Mars struck by massive meteor
sending ALH84001 into space

10 LINES
IN THE SEA

Red Bird of Paradise (*Paradisea rubra*),
collected by Alfred Russel Wallace for John Gould.
Ornithology Collection, Museum Victoria.

GREAT IDEAS HAVE no sense of occasion. They emerge, unexpectedly, from the most mundane observations and settings – on the back of a stained napkin in the university cafeteria or in the middle of a soporific presentation by an eminent visiting professor. Newton reputedly thought up the theory of gravity after watching apples fall in his mother's orchard (at least, according to Voltaire). The key to specific gravity was said to have occurred to Archimedes while he was having a relaxing bath.

Sometimes, however, the inspiration for a great theory is as grand and exotic as the theory itself. Charles Darwin's theory of evolution by natural selection was inspired by his observations of the strange creatures inhabiting the Galapagos Islands. Blood-supping finches and sea-going dragons perched on grim, black volcanic islands erupting from a sun-drenched azure sea – a romantic enough origin for any theory. Darwin's collaborator in developing the theory of natural selection, Alfred Russel Wallace, had a similarly exotic locale in dense jungles of South-East Asia. But 'the problem of the origin of species' was not the only theory to emerge from the fertile jungle or from Wallace's mind. For Wallace was puzzled not only by the origin of species, but also by their distribution around the globe. The results of his musings gave rise to the modern discipline of biogeography.

WALLACE AND DARWIN

It took Darwin seventeen years to put together the manuscript for *On the Origin of Species*, and it would have taken him even longer had it not been for a letter from an almost unknown zoological collector, Alfred Russel Wallace.

Wallace was a remarkable man. Unlike many scientists of his day, he had no private income and could not afford to attend university. In his youth he worked as a surveyor and then as a teacher, all the while seeking to improve his education and find a way to extend his explorations of the natural world. Using £100 he had saved as a surveyor, Wallace boarded a ship for Brazil. His travelling companion, Henry Bates, reported that it was their intention to 'make for ourselves a collection of objects, dispose of the duplicates in London to pay expenses, and gather facts, as Mr. Wallace expressed it in one of his letters, "toward solving the problem of the origin of species"'.

And solve it he did – while convalescing from malaria during his second major expedition to the Malay Archipelago. Like Darwin, Wallace was greatly impressed by the work of the English

PLATE 20. Trays of Wallace's insect specimens from Museum Victoria.

PLATE 21. The Chestnut Quail-Thrush (*Cinclosoma castanotus*) was traditionally classified in the same family as the European babblers, but molecular taxonomy places it clearly in its own Australo-Papuan family, the Pomatostomidae.

E. Gould.

clergyman and mathematician Thomas Malthus (1766–1834), who suggested that the exponential increase of the human population was prevented only by untimely deaths from disease, famine and wars. Wallace later recalled: 'there suddenly flashed upon me the idea of the survival of the fittest – that those individuals which every year are removed by these causes, – termed collectively the "struggle for existence" – must on average and in the long run be inferior in some one or more ways to those which managed to survive.'

Alfred Russel Wallace.
State Library of Victoria.

The more Wallace considered the consequences of this process, the more excited he became. 'In this way, every part of an animal's organization could be modified exactly as required, and in the very process of this modification the unmodified would die out, and thus the definite characters and the clear isolation of each new species would be explained. The more I thought it over the more I became convinced that I had at length found the long-sought-for law of nature that solved the problem of the origins of species.'

Wallace quickly constructed a paper and sent it to the only person he knew of who had an interest in the question – Charles Darwin.

Wallace's letter and paper stunned Darwin. His great idea was on the verge of being trumped. Strictly speaking, Wallace's completed manuscript had precedence over Darwin's years of work, since none of that was yet ready for publication. Darwin appealed to colleagues for advice. They suggested that Darwin and Wallace publish a joint paper, 'On the Tendency of Species to Form Varieties: and on the Perpetuation of Varieties and Species by Natural Means of Selection', outlining publicly for the first time the theory of evolution by natural selection.

Wallace was delighted by the resulting collaboration, even though it toned down his conclusions and gave prominence to Darwin's unpublished notes over his complete manuscript. For his part, Wallace seems to have regretted forcing Darwin to go public before he felt fully prepared. Certainly no animosity or recriminations were apparent between the two men. Wallace was one of Darwin's most ardent supporters and spokesmen, and their continued correspondence was one of the most productive partnerships in the biological sciences. Not only did Wallace spur Darwin to action over natural selection, but Darwin was pivotal in coalescing Wallace's thoughts on another seminal theory of the biological sciences – how to explain the distribution of species around the globe.

Wallace was a man of his times, and he lived in remarkable times indeed. The nineteenth century was characterised by a coincidence of great colonial and exploratory activity around the world, with enormous interest in the development of biological sciences. Certainly these activities were not unconnected. Biological theories were fuelled by the vast collections of strange and unusual creatures flooding into Europe from foreign lands, while advances in biology spurred museums and collectors to demand more and greater discoveries. Wallace played an active role in this process. Over his life he travelled extensively in Asia and South America. From the 'Far East' alone, he sent home 310 mammals, 100 reptiles, 8050 birds, 7500 shells, 13,100 butterflies and moths, 83,200 beetles and 13,400 other insects — more than 125,000 specimens in total (Plate 21).

Prior to the discovery of the New World, biology was a very local study. Few Europeans had any concept of the great diversity of life forms on other continents, and much of what they did know was a mix of myth, fantasy and occasional observations from untrained sailors and explorers. The exploration of new continents, however, dispelled this view.

In 1761 a French naturalist, Georges Buffon (1707–1780), observed that America shared no mammals at all with Europe (and that those in America were smaller and inferior). Nothing could have brought the notion of regional differences in species home more forcefully than the exploration of Australia in 1768–1771 by Joseph Banks under the captaincy of James Cook. Here was an entire continent swathed year-round in strange, sparsely clad eucalypts and acacias, filled with the riotous shrieks of parrots never seen before, and populated by strange mammals like the pouched marsupials and bird-like mammals. Not only were the species different to those familiar to Europeans, but (unlike America) many did not even belong to the same families of dogs, cats, deer and so on. By 1820 regional differences in plant life were formally recognised by the Swiss-born French botanist Augustin Candolle (1778–1841), who proposed twenty distinct botanical regions around the world. But no-one had applied these concepts to animals and there was little, if any, information on the subject in English.

Creationists saw no particular difficulty in these remarkable discoveries. Clearly God had simply populated different continents with their own unique sets of creatures. Wallace, familiar with the intricacies of animal distribution, was not to be fobbed off with such a simplistic answer. The creatures of the different continents might be quite different, but they also shared striking similarities. Why are rodents and bats so widespread, while marsupials are restricted to Australia and southern America? Wallace was a well-travelled man and was confronted on a daily basis with facts few of his English colleagues cared to imagine. He not only experienced the vast diversity of life, but was observant enough to notice patterns in that diversity – patterns which could not be explained by prevailing theories. But even a knowledge of these general inconsistencies could not have prepared Wallace for the remarkable state of affairs he found in the Malay Archipelago (present-day Indonesia, Malaysia, Brunei, the Philippines and East Timor). And the location of Wallace's moment of truth (at least in hindsight) was the inauspicious island of Bali and its nearby neighbour of Lombok, which Wallace visited only in order to get from Singapore to Celebes (now Sulawesi).

The schooner anchored off the north coast of Bali, and Wallace used the two-day stopover to explore. He was astonished by the highly cultivated nature of the land and the intricate systems of irrigation and rotation cropping. 'In so well-cultivated a country it was not to be expected that I could do much in natural history.' In any case, most of the bird species Wallace saw were much the same as those he knew from Java – the Asian Golden Weaver (*Ploceus hypoxanthus*), wagtail-thrushes, orioles and starlings. At the time Wallace was unaware that this was the easternmost point at which these birds occurred.

Then Wallace landed, with all his equipment, on the nearby island of Lombok amidst pounding surf on a black volcanic beach. Here he planned to devote himself to collecting with more rigour, while he awaited a boat to Macassar (Ujung Pandang on Sulawesi).

> Birds were plentiful and very interesting, and I now saw for the first time many Australian forms that are quite absent from the islands westward. Small white cockatoos were abundant, and their loud screams, conspicuous white colour, and pretty yellow crests, rendered them a very important feature in the landscape. This is the most westerly point on the globe where any of the family are to be found. Some small honey-suckers of the genus Ptilotis, and the strange mound-maker (*Megapodius gouldii*), are also here first met with on the traveler's journey eastward.

It was a pattern of division between Asian and Australian birds which Wallace was to record on other islands in his journeys through the region. But the division between Bali and Lombok was perhaps the most stark because of their proximity to one another – separated by a strait just 25 kilometres wide. While birds of both Australian and Asian families are found on both Bali and Lombok, Wallace felt that there was a significant increase in the proportion of Australian birds on Lombok compared to

Wallace illustrated his work on the geographical distribution of animals with plates entitled 'A forest in Borneo, with characteristic mammalia', showing both Asian and Australasian fauna (left) and 'Scene in New Guinea, with characteristic animals'.

Bali. Bali seemed more similar to its westerly neighbours of Java, Borneo and Sumatra than to its closer eastern neighbour of Lombok. In contrast, Wallace felt that Lombok shared more in common with its eastern neighbours of Sumbawa, Flores and Timor – even the more northerly Sulawesi – than with nearby Bali. The strait of Macassar between Borneo and Sulawesi, and the Lombok strait became known as the Wallace Line, highlighting a difference in faunal composition which Wallace felt had significant implications for understanding the geology of the area.

Wallace obviously discussed his observations with other scientists for in 1858, Philip Sclater (1829–1913) used Wallace's observations to suggest that division between regions dominated by 'Oriental' birds and regions dominated by Australian birds fell between Bali and Borneo to the west and Lombok and Sulawesi to the east. Wallace himself published his observations on birds in a brief paper in 1859 and, in the following year, extended his ideas to encompass mammals and insects as well. The important distinction between Sclater and Wallace was that Wallace was not simply interested in describing regional differences; he wanted to understand why such regional differences occurred. The only logical explanation to Wallace was a combination of evolutionary changes and geological history. He wrote, 'I believe the western part [of the Malay Archipelago] to be a separated portion of continental Asia, the eastern the fragmentary prolongation of a former Pacific continent.'

In part, Wallace's conclusion about the continental origins of the Malay islands was prompted by Darwin. Darwin had noticed that differences in the fauna of land masses were greatest where the water between them was deepest. When Wallace applied this idea to the Malay Archipelago, he found a near perfect match. The idea of rising and falling sea-levels was a relatively new one, promoted largely by Charles Lyell (1797–1875) in his three-volume *Principles of Geology*. Lyell's work argued for a gradually changing history of Earth's geology, rather than either a static one or a history punctuated by dramatic catastrophes (like the Biblical flood). Lyell did not apply his 'evolutionary' arguments to biological systems (indeed, until he was persuaded otherwise by Darwin, he supported the immutability of species). It remained for the likes of Darwin and Wallace to carry the implications of Lyell's geological theories through to living organisms.

Lyell felt that the distribution of plants and animals around the globe was largely influenced by two factors. The inherent ability of many species to disperse allowed some to spread, but their inability to adapt to new environments or compete with pre-existing species better suited to those environments constrained their spread. Extinction (essential for Lyell to explain the fossil record) was evidence of species' inability to adapt to environmental change. In this Lyell rejected Lamarck's transmutationist evolutionary theory which allowed for rapid adaptation. Darwin and Wallace's theory, of course, offers a middle ground between Lamarck and Lyell. Species cannot adapt to environmental change within one generation as Lamarck thought, but over the course of many generations (if environmental change is gradual) they can adapt. Extinction is still common, but not quite as inevitable as Lyell argued.

Wallace probably went to South-East Asia with Lyell's assumption that islands are largely populated by species which spread there by dispersal. But his observations did not fit this theory at all. To the contrary, Wallace was struck by the dissimilarity between some close island neighbours, such as Bali and Lombok. If dispersal was so important, these islands should have almost identical fauna.

For Wallace, the only feasible explanation for the pattern of similarities and differences between the islands of the Malay Archipelago was that similar islands had been connected by a land bridge more recently than those islands which showed the greatest divergence of species. Islands with a range of animals completely different from their neighbours' may never have been connected at all. An island's dissimilarity, in terms of its inhabitants, was a reflection of its long isolation. And this theory was supported by the depth of the waters surrounding the islands. The potential for land links between the islands can most clearly be seen on a map of the continental shelf surrounding each land mass. The deep channel between Bali and Lombok, for example, would have existed even when the oceans were at their lowest. The shallow sea between Borneo and Java, however, would have provided a land-link when the oceans were lower, explaining the greater abundance of similar families on both islands. Wallace's observation led him to formulate the basic principle of biogeography – that the distribution of plants and animals may reveal the geological age and past history of land masses which, in Wallace's time, could not be investigated directly. Wallace's two contributions on the subject, *Geographical Distribution of Animals* and *Island Life*, were among the first books available in English on biogeography.

TOO MANY LINES?

The Wallace Line became a source of great contention. Biologists flocked to see the near magical faunal break, some even suggesting that by crossing the strait between Bali and Lombok one could step back in time 100 million years from the modern advanced fauna of Eurasia to the supposedly ancient primitive creatures of Australia. Other scientists were less convinced, and numerous alternative lines were proposed.

In 1866 Andrew Murray (1812–1878) suggested that the line actually ran to the west of Bali. In 1896, the curator at the British Museum of Natural History and expert on Australasian mammals, Richard Lydekker (1849–1915), suggested that the greatest change in the proportion of Australian to Asian fauna actually occurred on the opposite side of the Malay Archipelago, around the islands located on the Australo-Papuan (Sahul) continental shelf. In 1899, Sclater (aided by his son) revised his earlier agreement with Wallace and proposed that Sulawesi be

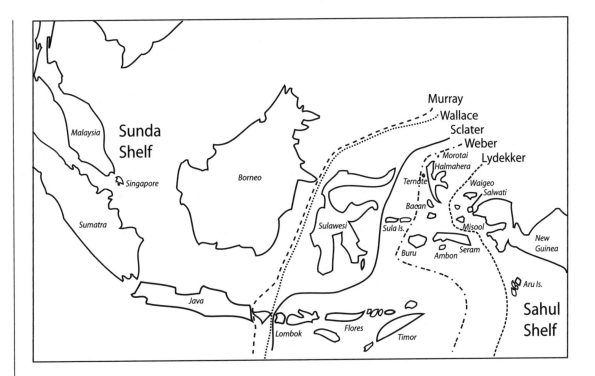

counted on the Asian rather than the Australian side. And in 1902, Max Weber (1852–1937) proposed a line that ran roughly through the middle of the island chain, based on the distribution of freshwater fish.

Most of these biogeographic lines have sunk into obscurity, with only the Wallace Line being remembered for its historical, if not scientific, significance. When evolutionary biologist Ernst Mayr reviewed the subject in the 1944, he supported Weber's Line because it falls roughly along the point at which the balance between Australian and Asian families is equal. Others have castigated modern biogeographers for being too eager to find a particular line so that they can draw their maps with different colours, rather than allowing a zone of overlap. Mayr proposed a line of 'faunal balance', falling along the area where there are roughly 50 per cent Australian and 50 per cent Asian species for all taxonomic groups.

In 1977 George Simpson concluded that there were simply 'too many lines' and that the whole area known as 'Wallacea' between Wallace's Line and Lydekker's Line should simply be regarded as a zone of transition between the Asian fauna and the Australian fauna.

After Wallace had proposed a location for the division between Asia and Australasia, other scientists suggested alternatives.

The Wallace Line has fallen into disrepute largely through its own success. Few modern biogeographers can imagine a world where dispersal is the only explanation for animal distribution and where no-one considers geology to be an important factor. Modern biogeographic boundaries are proposed in order to test theories about animal distribution and geological history. Wallace's biogeographic boundary was proposed as a theory that there was a relationship between animal distribution and geological history. The scientific validity of Wallace's conclusions depends not on whether they support modern biogeographic theories, but on whether the facts and observations that Wallace made supported the theories he proposed.

Wallace was not interested in merely placing boundaries around faunal regions. Such specific boundaries will differ for taxonomic groups which have different evolutionary histories, environmental requirements and barriers to dispersal. The very principle that Wallace was proposing – that the current distribution of animals reflects geological history – predicts varying biogeographic boundaries for creatures of dramatically different antiquity.

Although Wallace acknowledged that there was actually a transition in the proportion of Asian to Australian species from the west to the east of the Malay Archipelago, he felt it was not the smooth, even transition one would expect if dispersal alone accounted for the distribution of species from each mainland continent. Wallace felt there was an unexpected blip in the transition from Asian to Australian species from Borneo to Sulawesi and from Bali to Lombok. This blip is an echo of geological history.

The theory of dispersal predicts that there will be far fewer species on islands than on the mainland. Smaller land masses also harbour fewer species than larger land masses. The greatly reduced number of species on the islands of Wallacea has led many modern researchers to suggest that the Wallace Line is not a biogeographic boundary at all, but merely separates the rich continental fauna of Asia from the less diverse island fauna.

Wallace was well aware that these intermediate islands contained far fewer species than those closer to mainland Asia. But the total number of species makes no difference to the proportion of Asian or Australian birds on each island. Wallace himself recorded that Sulawesi only had 220 species of birds compared to the 386 species found on nearby Borneo, but he also noted that, while nearly all of Borneo's birds are Asian in origin, only three-quarters of Sulawesi's birds are Asian. It is highly unlikely that Wallace thought that the reduced abundance of species on the intermediate islands translated to a change in the proportion of Asian and Australian species.

So did Wallace have the data to support his conclusions? Unfortunately, Wallace's papers do not record the actual proportions of Australian and

Sulphur-crested cockatoo (*Myctolophus galeristus*, now known as *Cacatua galerita*). Some modern scientists have suggested that Wallace was misled into placing his line too far west by the noisy and conspicuous 'Australian' birds he first noticed on Lombok. J. J. Halley (1871).

Asian species he observed for every island – such statistical evidence was not a requirement for scientific papers of the day. But Wallace's papers are not the only legacy from which we can judge his science.

Wallace was a superb collector and preparator of specimens. Specimens that he collected fill the drawers of museums around the world, their exceptionally neat preparation a testimony to the skill with which he tackled his work under trying conditions. Even today, a Wallace specimen stands out among specimens prepared by less-skilled taxidermists, a remarkable feat considering the environment in which he worked. Wallace describes how his collecting operations in Lombok were more than usually difficult:

> One small room had to serve for eating, sleeping and working, for storehouse and dissecting-room; in it were no shelves, cupboards, chairs or tables; ants swarmed in every part of it, and dogs, cats and fowls entered it at pleasure. Besides this it was the parlour and reception-room of my host and I was obliged to consult his convenience and that of the numerous guests who visited us. My principal piece of furniture was a box, which served me as a dining table, a seat while skinning birds, and as the receptacle of the birds when skinned and dried. To keep them free from ants we borrowed, with some difficulty, an old bench, the four legs of which being placed in cocoa-nut shells filled with water kept us tolerably fee from these pests. The box and bench were however literally the only places where anything could be put away, and they were generally well occupied by two insect boxes and about a hundred birds' skins in process of drying. It may therefore be easily conceived that when anything bulky or out of the common way was collected, the question 'Where is it to be put?' was rather a difficult one to answer. All animal substances, moreover, require some time to dry thoroughly, emit a very disagreeable odour while doing so, and are particularly attractive to ants, flies, dogs, rats, cats and other vermin, called for especial cautions and constant supervision, which under the circumstances above described were impossible.

Wallace goes on to excuse himself for collecting less than he would have liked to – a mere forty-three specimens for each day he was in the area! These specimens are now scattered throughout the museums of the world as well as private collections. Wallace specimens were highly sought after by museums and collectors alike. Frederick McCoy actively sought Wallace specimens by urging the entomologist at the British Museum to encourage his friend Wallace to send his specimens to Australia, and not just Britain. McCoy must have met with some success, because careful exploration of the ornithology collection of Museum Victoria by Ian McAllan and Rory O'Brien has revealed at least forty-five specimens collected by Wallace.

The way in which collections are catalogued usually makes it difficult to locate specimens made by a particular collector – unless the specimens were sold or donated to a museum in a bulk lot, in which case they will tend to be entered into an acquisitions catalogue sequentially. And this is precisely what happened when the British Museum of Natural History acquired 2474 bird specimens that Wallace had collected in the Malay Archipelago. The acquisitions register for 1871 records the species and location for each specimen. This provides us with a sample of the original 'data' with which Wallace formulated his line.

THE ORIGIN OF BIRDS

When Wallace allocated his bird specimens to either Oriental or Australian groups, he based his decision on a detailed knowledge of the distribution of species. An Oriental species is one that belongs to a family where most members are found in Asia. When Mayr reassessed the Wallace Line, he too used 'personal knowledge' to argue whether a family was Oriental or Australian, claiming that 'a specialist of a given group usually has no difficulties in deciding which species are Indo-Malayan and which Australian'. In some senses, these analyses are somewhat circular – species that are commonly found on the Austro-Papuan side belong to families that have more species on the Austro-Papuan continent. But neither Wallace nor Mayr had any more objective basis on which to determine the origin of the birds of the Malay Archipelago. Avian fossils have provided more recent analyses of the Wallace Line (such as Alan Keast's) with a stronger basis on which to found these distinctions. But avian fossils are rare in Australia and there is insufficient material from which to draw any firm conclusions about where different bird families originated (Plate 21).

Molecular genetics has revolutionised our understanding of the evolutionary relationships between song-birds, and thrown many of the traditional taxonomies into disarray, particularly in Australia. Many Australian birds were identified by European scientists as variants of European birds, particularly the small songbirds which superficially look similar. The Australian Flame Robin (*Petroica phoenicea*) was named after the European Robin (*Erithacus rubeculus*). The White-browed Tree-creeper (*Climacteris affinis*) was thought to be a southern version of the European Creeper (*Sitta europea*). Flycatchers, honeyeaters and wrens were all assigned names and relationships based on their similarity to European species. With this system of nomenclature came the assumption that Australian birds were derived from the European species, in waves of immigration which swept down through Asia and into Australia about 10–15 million years ago.

In Wallace's time it was assumed that the oceans and continents were fixed in place, and thus the only feasible explanation was that Europe and Asia were the main 'workshops' for the origin of most vertebrate species. Australia had been populated by an early wave of Eurasian birds. These primitive species survived in Australia but were later replaced by more advanced birds in Europe and Asia. It was not until the theory of continental drift was adopted that the notion of avian waves of immigration from north to south was exploded. Continental drift places Australia in splendid isolation until 20–30 million years ago when it collided with Asia, yet the oldest bird fossils in Australia are 120 million years old and many modern groups date from 30–40 million years ago.

Molecular patterns suggest that Australian and Papuan songbirds can be roughly divided into two groups – the Corvi (honeyeaters, wrens, robins, Australian warblers, logrunners, treecreepers and lyrebirds) and the Muscicapae (including flycatchers, whistlers, magpies, cuckoo-shrikes, orioles and fantails). By pinpointing the molecular 'age' of these birds (or the time at which they began to speciate), Charles Sibley and Jon Alquist suggest that the Corvi diverged from other birds about 58–60 million years ago, when Australia lay 3000 kilometres south of Asia.

When Australia finally came into close contact with Asia, the islands of the Malay Archipelago formed stepping stones for the dispersal of these Australian species into Asia and for species found in Asia to move into Australia. The Australian corvids radiated out into Europe and America from Asia as crows and jays. Asian birds (which may have originated locally or in Africa) reciprocated, with the ancestors of the Muscicapae moving into Australia.

This molecular work gives us a far clearer idea of which of Wallace's birds are actually Australian in origin and which are Asian. Some groups, whose representatives have been found in both Eurasia and Australia, have been split by the molecular research into separate families; for instance, the kingfishers are now divided into the Daleconidae or wood kingfishers of Australian origin and the Alcendinidae of Eurasian or African origin. Similarly, DNA studies have clearly divided the Australian warblers (Pardalotidae) from their Eurasian look-alikes (Sylviidae), and the Australian robins (Petroicidae) from the Old World flycatchers (Muscicapidae). Prior to these molecular studies, such groups would have been classified as 'cosmopolitan', blurring the distinction between Australian and Asian bird groups.

Although bird specimens form a small part of Wallace's complete collections, they seem to have been a key element in his formulation of the Wallace Line. His early observations in Bali and Lombok were of birds. His descriptions commonly refer to birds and his first papers on the subject rely heavily on the observation of bird distributions.

Once we know which birds are from Australian families and which are from Asian families, we can calculate the overall proportion of Asian species Wallace collected from each island (excluding cosmopolitan species whose families are equally abundant on both continents, such as many seabirds and waterfowl). From the specimens in the British Museum and Museum Victoria, we have twenty islands from which Wallace collected enough specimens to include in the analysis. Unfortunately Wallace did not collect enough specimens from some islands to include in these calculations. From Bali, for example, he collected only nine different species. But for the remaining islands, the proportion of Asian species gradually declines the further east you travel. Of the species Wallace collected in Borneo, 72 per cent were from Asian families, while just 2 per cent of the species from Waigeo on the western tip of New Guinea were Asian. But is this transition entirely smooth, or is there a point at which the proportion of Asian to Australian species changes unexpectedly?

Wallace's data conform perfectly to his prediction that the modern distribution of species reflects the geological history of the land masses. Modern geological knowledge suggests that the islands on the Sunda Shelf were originally part of mainland Asia before becoming separate islands (off and on) over the past 30 million years. Land bridges remained as recently as one million years ago, when the sea fell during the last Ice Age. Similarly, the islands on the Sahul Shelf were originally connected to New Guinea as well as Australia and have periodically been reconnected until recent geological times. The islands in the middle, including Sulawesi, Lombok, Timor, Seram and Halmahera, have a far more isolated

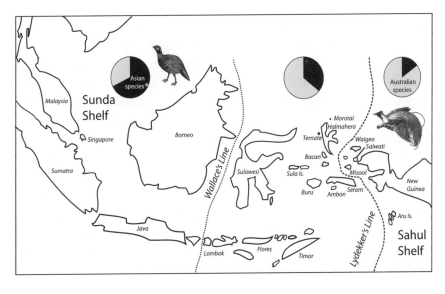

The proportion of Asian birds collected by Wallace on the islands west of the Wallace Line (on the Sunda Shelf) is much greater than the proportion of Asian birds from islands east of the Wallace Line. However, the proportion declines noticeably again for the islands on the Sahul Shelf (east of Lydekker's Line).

biogeographic history, with similarities to Asia and Australia primarily derived from dispersal rather than physical connection.

Wallace's line may or may not be as strong for taxonomic groups other than birds. But the fact remains that the Wallace Line describes precisely what he said it did – a sharp change in the proportion of Asian to Australian species. In addition, Wallace's argument that the geographic patterns of animal distributions reflect the former geological history of the lands they occupy has achieved such universal acceptance that modern biogeographers tend to forget that this was the primary purpose of Wallace's argument. Far from being tarnished by inaccuracy, the value of the Wallace Line has been overshadowed by the very success of the theory it was put forward to support.

The more interesting question is why didn't Wallace notice Lydekker's Line? Wallace spent a considerable period of time based on Ternate, a small island on the western side of Halmahera which lies right on the edge of Lydekker's Line, collecting extensively from New Guinea and surrounding islands, particularly the popular and spectacular birds of paradise. But a closer inspection of Wallace's accounts of his travels reveals that most of these collecting trips across Lydekker's Line were undertaken by his assistant. Wallace during this period spent an increasing amount of time recovering from malaria. The vast majority of Wallace's sojourn in the Malay Archipelago was spent either on the 'Asian' islands or on the intermediate islands. Wallace did not collect all his specimens personally but, as was common among collectors of the time, utilised local networks. If Wallace made greater use of such networks in the east than he did in the west, his opportunities for making a direct comparison between the two sides of Lydekker's Line would have been greatly reduced. Even among the birds he did collect, almost 500 come from the New Guinean islands, while over 800 come from the Asian islands and over 1000 from the islands in between.

More importantly, Wallace had no reason to suspect the existence of a second line. He was after all, primarily interested in demonstrating the existence of differing continental origins for the islands, not a completely isolated 'oceanic' origin for some of them. And if Wallace had been prepared to accept the hints his faunal evidence was suggesting, of an African connection, what reasonable geological theory could he possibly have used to explain it? Wallace believed that continents had risen and fallen, eroded and erupted over the course of Earth's history. More adventurous geological theories risked ridicule and, more importantly, were not supported by the geological information available at the time. It would be many years before geology caught up to biogeography and provided an explanation for some of the mysteries Wallace had uncovered.

11 SHIFTING CONTINENTS

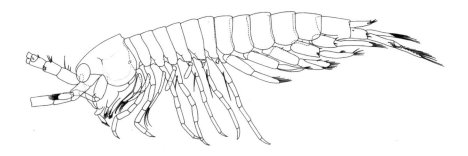

A rare *Spelaeogriphacea* crustacean from underground waters in the Pilbara Desert region of Western Australia, identified by Gary Poore, Senior Curator in Crustacea, Museum Victoria. The only other living members of this ancient order are found in a cave stream on Table Mountain in South Africa and in a cave lake in the Mato Grosso of Brazil, adding to the growing list of fauna with Gondwanan origin.
Marine Invertebrate Collection, Musuem Victoria.

GEOLOGY CAN ALMOST be seen as underpinning biology. The rocks determine the soil and fertility, which in turn determines the plant communities that develop and the animals that the plants support. The entire ecosystem is built from the nutrients contained in the rocks that lie beneath it. Intellectually, too, the science of geology can sometimes be seen as underpinning many key developments in the biological sciences. Geological research has often laid the foundations for some of the most significant discoveries in biology, particularly evolutionary theory. It was the work of the geologist Charles Lyell (1797–1875), for example, which laid the foundations for Charles Darwin's theory of natural selection.

Lyell's influential work appeared at a time of great debate in geological theory. Increasingly, research into the nature of the Earth was challenging the view that we live in an unchanging world which remains just as God designed it in 4004 BC (as calculated by Bishop James Ussher, 1581–1656). The rocks reveal unambiguous evidence of environmental change, of oceans come and gone, eroded mountains, exploded volcanoes and ancient riverbeds. There were many and various explanations for this fossilised history. The two which have received the greatest subsequent attention were the 'catastrophic' theory led by the French palaeontologist Georges Cuvier (1769–1832), and the 'uniformitarian' theory formulated by the Scottish geologist James Hutton (1726–1797) and promulgated by Charles Lyell. Today, elements of both arguments are accepted as valid explanations for the geological history of the Earth, but at the time each of these geological approaches had significant implications for the development of evolutionary theories.

A catastrophic view of geological history allows for sudden and dramatic environmental change, such as floods on a scale greater than that seen in living history. Cuvier believed that species were fixed in nature and did not alter in response to environmental change. Environmental catastrophes could explain how species which appeared in the fossil record at certain times were no longer present on Earth. Fossils are characterised by a series of dramatic changes (or 'revolutions' as Cuvier termed them) in the suite of species present. Cuvier identified six such geological epochs (which English geologists, conveniently, loosely translated into the six 'days' of the book of Genesis). The periods of change between these epochs are mass extinction events. The most famous mass extinction featured the demise of the dinosaurs – between the Cretaceous and Tertiary periods. But actually the early Permian–Triassic extinction (245 million years ago) was far more dramatic, with up to 96 per cent of marine fossil species being wiped out. Cuvier felt that these dramatic alterations in the patterns of life present on Earth could only be explained

Charles Lyell (1797–1875), famous for advocating a uniformitarian view of geology and for refuting the works of Lamarck (but also introducing them to an English-speaking audience).

PLATE 22. Cassowaries are members of the large family of flightless birds known as ratites.
H. Richter.

PLATE 23. Viking 2 panorama of the Martian surface.
Courtesy of NASA and the National Space Science Data Center.

PLATE 24. Stromatolites at Shark Bay in Western Australia.
Calcium Carbonates Ltd.

in terms of catastrophic geological events. The Chilxcub meteor strike, currently argued to be a major contributor to the Cretaceous–Tertiary extinctions 65 million years ago, certainly conforms to the theory of catastrophism, although palaeontologists now think that the other extinction events probably involved more gradual processes.

Individual fossils also suggested to Cuvier that environmental change was rapid. Many fossils appear to reveal a life cut short unexpectedly. A fossil Mastodon (on which Cuvier was an expert) was found with food still in its mouth. Cuvier argued that certain marine fossils had been preserved *in situ*, not swept down and collected from somewhere else by tidal forces. These fossilisation events, rather than being a curious feature of fossilisation itself, were seen by Cuvier as evidence that death (and extinction) came rapidly. Both species and individuals seemed to have been struck down with Biblical might.

Cuvier spent most of his working life studying comparative anatomy at the Muséum d'Histoire Naturelle in Paris. But despite his huge influence, not all of his colleagues agreed with either his catastrophic view of the Earth's past, or his insistence on the fixity of species. Among Cuvier's senior colleagues, Jean-Baptiste de Lamarck (1744–1829) stuck doggedly to his belief in transmutation of species through the inheritance of acquired traits. No amount of ridicule from his colleagues, or even from Napoleon himself, could deter the seemingly sensitive and frail Lamarck from his central beliefs. Where Cuvier the anatomist was particular, precise and empirical, Lamarck (originally a botanist but happy to dabble in a variety of sciences) was philosophical, eccentric and eclectic. Nor was Cuvier's one-time mentor and friend, Geoffroy Saint-Hilaire (1722–1844), convinced by Cuvier's unwillingness to engage in broader speculation. The French scientific establishment dominated biology for decades, but across the English Channel rival theories were developing.

James Hutton, the Scottish pioneer of uniformitarianism, had been unconvinced by the theories that explained current geological patterns by processes which no longer occur. Dramatic floods and violent volcanic eruptions of a magnitude not observed in living memory seemed unnecessary when, over a longer time scale, the known gradual processes of erosion, sedimentation and volcanism could sufficiently explain geological patterns. Lyell developed this theory into his influential *Principles of Geology*, which presented a history of the world in terms of gradual and slow geological change, opposing both Cuvier's catastrophism and Lamarck's transmutation.

Both Darwin and Wallace were greatly influenced by Lyell's ideas and extended this 'evolutionary' (rather than revolutionary) geological framework to the biological world as well. Unlike Lamarck's transmutationist theory, natural selection requires an extremely long time scale to operate. Only given enough time can gradual change, such as that accumulated over successive generations, be sufficient to provide enough variety for evolution to operate. For natural selection, time is of the essence, and lots of it.

Despite their differences, Cuvier, Lamarck, Darwin and Wallace all recognised the changing nature of the history of life on Earth. This recognition both fuelled and was fuelled by the increasing appreciation that the Earth was considerably older than previously thought. Darwin wrote in *On the Origin of Species* that 'the belief that species were immutable productions was almost unavoidable as long as the history of the world was thought to be of short duration'.

Alterations in geological theories promoted the development of evolutionary theory in some areas, but the limitations of geological theories held biology back in others. While Lyell's formulation of a constantly and gradually changing environment clearly influenced Wallace when he formulated his theory of natural selection, the lack of a rational geological explanation for Wallace's observations of animal distribution in the Malay Archipelago restricted his ability to formulate a clear biogeographic principle. Based on some of his observations, Wallace was certain that the distribution of animals across islands would reveal something of the hidden geological history of those lands. But the animals of some islands, such as Sulawesi, defied a sensible geological explanation. Here was an island clearly pinned between Australia and Asia, whose biological relations sometimes seemed closer to Africa. No geological theory of the time (other than a far-fetched land-link of Lemuria between Madagascar, India and South-East Asia) was up to the task of explaining that anomaly, and Wallace was left puzzled.

Just as developments in geology had once provided great impetus for the development of paradigm-shifting theories in biology, now biology provided the momentum for a paradigm shift in geology. Geologists had often bewailed the careless abandon with which biologists threw up land bridges and sank continents in an effort to explain the inexplicable patterns of plant and animal distribution. Arthur P. Coleman, president of the Geological Society of America, complained in 1916 of 'the great recklessness in rearranging land and sea [for] the convenience of a running bird, or of a marsupial afraid to wet its feet'. But faced with creatures which had clearly been isolated for millennia and whose nearest relatives were often located across thousands of miles of ocean in South America or Africa or even India, biologists could often find little in prevailing geological theories to help them explain their strange distributions.

The Temple of Serapis, illustrated in Lyell's *Principles of Geology* (1830), provided a well-known example of changing sea-levels during historical times, having been built above sea level and then submerged.

It was not until 1912 (more than fifty years after Wallace first arrived in the Malay Archipelago) that Alfred Wegener (1880–1930) put forward his theory of continental drift, which began to explain confusing patterns like Sulawesi. At the beginning of the nineteenth century, geological arguments still centred on issues of permanence and change. The antiquity of the Earth was now undisputed (although not resolved in detail) and alterations to the Earth's surface were acknowledged, but there was still controversy over the extent of these changes. Catastrophism had largely fallen from favour, leaving varying forms of uniformitarianism to argue the field. Some geologists held that the continents and oceans were largely stable, with temporary oceanic incursions and upheavals being responsible for changes in the geological and fossil record. Other geologists opted for a more dynamic process of change, arguing that physical changes were due to a gradual contraction of the Earth – this school of thought had a stronger inheritance from the catastrophists.

Neither school was prepared for the dramatic suggestion of Wegener that the continents drifted about on the surface of the Earth, colliding and retreating from one another, driven by some kind of lunar or rotational force. Under Wegener's model, the continental land masses had once been joined in a single supercontinent of Pangea, which subsequently split into the northern land mass of Laurasia and the southern Gondwana. Later movements divided Gondwana into Antarctica, Australia, South America, and Africa, with India moving north to rejoin the northern land mass.

Wegener drew on a broad array of geological evidence to support his arguments. Most obvious to the non-specialist is the jigsaw-like fit between separate continents like the west coast of Africa and the east coast of South America (the fit is even more conspicuous when the edges of the continental shelves are compared). But the distribution of plants and animals was also important to arguments about Wegener's theory. Fossil seed ferns (*Glossopteris*), for example, are found across South America, southern Africa, India and Australia – providing support for Wegener's reconstruction of an original Gondwanan continent. The very distinctiveness of Australia's fauna, as well as its similarity to some South American species, was a key illustration in Wegener's 1921 publication of *The Origin of Continents and Oceans*.

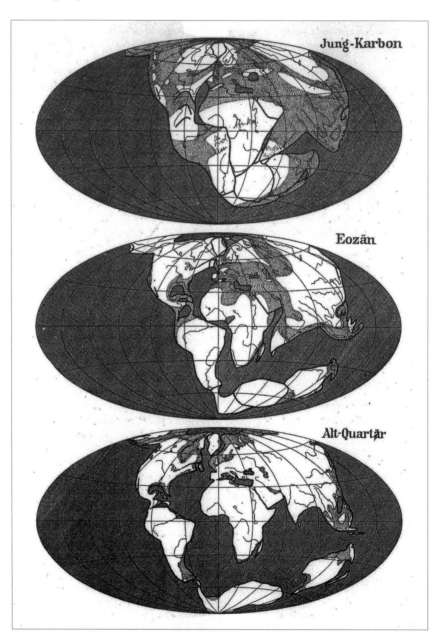

Map reconstructions at three stages of the earth's history illustrating tectonic drift theory.
A. Wegener

Wegener also drew upon Wallace's problematic South-East Asian data. Wallace's observation suggested to him that some of the islands in the Malay Archipelago had once been connected to India and Africa. All three shared characteristic fauna, most notably members of the *Prosimii* group – including lemurs, bushbabies and lorises. Prosimians (which are primates, along with monkeys and apes) are found in small numbers in Africa and Asia. By far the majority of prosimians are African, or more particularly, Madagascan. The true lemurs (Lemuridae) are only found in Madagascar, as are the dwarf or mouse lemurs (Cheirogaleidae), the Indri and Sifakas (Indriidae) and the strange long-fingered Aye-Aye (Daubentoniidae). The loris family (Lorisidae) are also primarily African – specifically the bushbabies and pottos. But the lorises themselves are Asian. The Slender Loris (*Loris tardigradus*) is found in India and Sri Lanka, while the Slow Loris (*Nycticebus coucang*) is found from Vietnam to Borneo. The final family of prosimians, the tarsiers (Tarsiidae) are found only on the isolated offshore islands of South-East Asia. The Philippine Tarsier (*Tarsius syrichta*) is found on a few islands in the southern Philippines. The Western Tarsier (*T. bancanus*) is found only on Borneo, Bangka and southern Sumatra. And the Spectral Tarsier (*T. spectrum*) is found only on the islands of Sulawesi, Great Sangihe and Peleng. Only a theory of continental drift, which saw India and some of the Malay Archipelago break off from, or intersect with, a larger continental mass including Africa before colliding into Asia, could explain such strange biogeographical patterns.

Sulawesi's strange Babirusa pig (*Babyrousa babyrussa*) reminded Wallace of African warthogs, and other animals, insects and birds also seemed to indicate an African link. Wallace had no geological evidence that could explain this. Remarkably, modern evidence suggests these central islands do have exceedingly ancient 'African' origins.

Drift theory not only explained some of the anomalies Wallace had puzzled over, but also provided theoretical backing for his central argument of a disjunction between Australian and Asian fauna. Wallace assumed that animals simply showed a remarkable ineptitude for crossing water barriers, even of a mere 25 kilometres (as between the islands of Bali and Lombok). Drift theory increased the physical dimensions of these barriers by suggesting that, in the past, the distance between 'Australian' islands and their 'Asian' counterparts had been considerably greater.

The battle of the theory of continental drift is one of the classic tales in the history of science. Wegener did not live to see his work vindicated, dying in the pursuit of science on the ice-cap of Greenland. Few geologists considered drift theory seriously until the mid-1950s, when new 'drift' theories and new geological evidence, particularly from the ocean floor, caused a resurgence of interest in the subject. Many scientists from the southern hemisphere took a particular interest in drift theory (or at least attacked it less vehemently than their northern hemisphere associates). The southern hemisphere contains many anomalies that were inadequately explained by prevailing geological theories. An anonymous

Lorises are a member of the Prosimii group (including lemurs, bushbabies and lorises), whose distribution across central Africa and southern Asia has long puzzled biogeographers.

geologist (quoted by Homer Le Grand) once quipped that 'most of earth's bigger scars, sutures and dimples etc. were more visible on her bottom, and the people from down under were quicker to see them'.

Throughout their respective histories, biology and geology have leapfrogged each other in relation to biogeographic theories. Sometimes, as in Wallace's time and in the development of Wegener's continental drift theory, biology is ahead of geology. Geological research in the nineteenth century was unable to investigate directly the origin of land masses, and Wallace felt that the distribution of animals could provide an insight into hidden geological processes. As geology has improved, assumptions about the history of animal distributions are dictated by the prevailing geological knowledge. But there remain anomalies between what geology tells us and what biology suggests.

Understanding how animals disperse is pivotal to understanding the relationship between biology and geology. Before Wallace's pioneering work on the subject, it was thought that terrestrial animals dispersed readily across oceans. Wallace, however, argued that animals were actually surprisingly bad at dispersing across water. In the absence of known geological processes, there are two ways in which animals can move from one island to another. The first is through dispersal – accidental spread of animals attached to flotsam and jetsam, creatures which are blown off-course by storms or arrive attached to birds and other travellers. Truly oceanic islands, such as volcanic Hawaii which has erupted from the ocean floor over the past few millennia, are colonised in such a manner.

A FLYING KIWI

The biogeographic patterns of southern hemisphere animals continue to provide challenges for current geological theories. Both kiwis and moas are members of a large family of flightless birds, known as ratites, distributed throughout the southern hemisphere. Australia has emus and cassowaries (pictured above), Africa has ostriches, South America has rheas and tinamous, while Madagascar had the elephant-bird and New Zealand has kiwis and, until recently, moas.

All of these birds are descended from a flying ancestor similar to the small, ground-dwelling South American tinamous. If this common ancestry is correct, the ratites were assumed to have evolved biogeographically. Thus the South American rheas were assumed to be most closely related to African ostriches. The New Zealand kiwis and moas were thought to be related more closely to each other than to the Austro-Papuan cassowaries (Plate 22). However, using DNA from extinct moa museum specimens, geneticist Alan Cooper demonstrated that the moas are actually quite distinct from their compatriots, the kiwis. Kiwis are more closely related to the emu and cassowaries. This remarkable discovery suggests that ratites colonised New Zealand twice, first resulting in moas and later resulting in kiwis.

The degree of genetic similarity between the moas and the Austro-Papuan ratites suggests that they shared a common ancestor about 82 million years ago – about the time New Zealand separated from Australia. But the kiwis share a much closer ancestor with their Australian cousins – about 68 million years ago. Either the ancestral ratite was a better-than-expected flier, or there has been a relatively recent link between New Zealand and Australia. Perhaps to the despair of geologists, the researchers suggested a possible land link between New Zealand and Australia via New Caledonia.

The study of other extinct New Zealand birds provides additional evidence for the latter scenario. The Piopio (*Turnagra capensis*) has not been seen in the wild since 1908 and is known only from museum specimens. Surprisingly, molecular studies revealed it to be most closely related to the Austro-Papuan bowerbirds, with whom it probably shared a common ancestor 27 million years ago. Apart from this single anomaly, bowerbirds are entirely restricted to regions once connected by land (such as New Guinea and Australia). Either the ancestral bowerbird was a better disperser than we thought, or there was a more recent land link between Australia and New Zealand than was previously thought.

The use of molecular evolution to 'time' when species diverged from one another is recent and contentious. Mutations certainly accrue in a species' genetic material at a relatively constant rate, but synchronising that rate to reflect a comparable geological time scale is difficult. Mutation rates are not the same across species either. Different life history strategies, particularly longevity and generation times, speed up or slow down the rate at which mutations accumulate. Molecular clocks need to be synchronised not only with geology, but also with each other. Irrespective of the timing of these events, however, the molecular relationships between these birds suggests a more complex biogeographic pattern than would be expected from our knowledge of geology and how ratites or bowerbirds are thought to have dispersed.

The second form of movement is known as vicariance – the spread of animals across land bridges or, in the case of marine organisms, along relatively shallow underwater ridges which are then cut off from their place of origin by changes in climate or geology. This dispersal can be seen as an extension of the animal's natural range, rather than an accidental outposting on an unexpected habitat. Prior to Wallace, most island animals were thought to have arrived accidentally, by dispersal. Wallace however, argued that most such animals had actually arrived by vicariance, and that when such land bridges were removed, the level of animal colonisation dropped dramatically. Truly oceanic islands, like those of Hawaii, Fiji, Tonga and the Galapagos, have relatively few terrestrial mammals.

The theory Wallace proposed over 100 years ago in a malaria-ridden tropical jungle applies just as much to tiny sea stars clinging to the shores

Wegener used evidence from around the globe, including sea depths surrounding New Guinea, to support his theory of tectonic plate movement. A. Wegener (1924).

Part 4 VISIONS OF NEW WORLDS

of a salt-lashed and frozen oceanic island. The distribution of life on Earth may echo the geological forces that have shaped the Earth's surface, but our understanding of geology and biology is limited by our understanding of the history and behaviour of the plants and animals populating the diverse regions of the globe. Our efforts to understand the distribution of species in time and space has driven two of the most significant theories in biology and geology of our times.

And now this same question looks set to drive a new paradigm-breaking phase in the development of the natural sciences. Just as global exploration in the seventeenth and eighteenth centuries led to an intellectual boom in natural sciences, today's space exploration looks set to expand our horizons to question not only the distribution and origin of life on Earth but, perhaps, across the whole universe.

MACQUARIE ISLAND

Macquarie Island emerged from the subantarctic Southern Ocean 27 million years ago on the Macquarie Ridge south of New Zealand, between the Pacific and Australian tectonic plates. It lies 1500 kilometres off south-eastern Australia and has never been attached to another land mass or even been particularly close to any other land mass. Not surprisingly, Macquarie Island has relatively few native species compared to islands closer to larger land masses. Elephant seals (*Mirounga leonina*) and a few species of fur seal share the island with millions of penguins, gannets and other seabirds. Among the invertebrates, a team from Museum Victoria found just seventeen species in the freshwater streams of the island. Given Macquarie Island's long isolation surrounded by salt water, freshwater fauna would be expected to have few affinities to species on the continental land masses.

Most of Macquarie Island's stream invertebrates belonged to typically marine families that can adapt to freshwater, such as marine tubeworms (turbellarians) and isopods, or were flying insects, such as craneflies, which breed in water. But the most abundant species in Macquarie Island's streams are freshwater oligochaetes worms. How they arrived on Macquarie Island (and other isolated oceanic islands) is a mystery.

Another Macquarie Island mystery is posed by *Pseudoboeckella brevicaudata*, a shrimp-like copepod with a circumpolar distribution on most islands and all southern continents, yet it cannot tolerate salt

Macquarie Island Sea Star (*Crossaster multispinus*).

water. Another species, *Astacopsidrilus*, has previously been recorded only in New South Wales and Queensland. Its eggs may have arrived on the feet of visiting Australian seabirds or ducks which seek out these islands in times of drought on the Australian mainland.

Such speculation as to the origin of aquatic organisms on subantarctic islands focuses on theories of dispersal. Most of the shallow-water marine invertebrates, for example, were assumed to have arrived on these islands from continents to the west by dispersing on the West Wind Drift. Large clumps of kelp (*Macrocystis pyrifera*) drift from mainland coasts around the Southern Ocean, presumably taking with them large numbers of resident creatures. But what real evidence is there that animals have dispersed in this way to isolated outcrops like Macquarie Island? Tim O'Hara, from Museum Victoria, investigated this question by looking at the origins of Macquarie Island's echinoderms.

Echinoderms include the sea stars, sea urchins, brittle stars, feather stars and sea cucumbers. Of the fifty species known to occur in the coastal waters of Macquarie Island, O'Hara found that relatively few species had their nearest relatives on land masses to the west (as would be expected from West Wind Drift). Dispersal via the West Wind Drift clearly cannot account for the arrival of species from the south or north or east. Interestingly, those species that did originate on land masses to the west had strong tubefeet, were from rocky shallow-water areas where kelp occurs and brooded their eggs internally, making them good candidates for dispersal on kelp.

The majority of species, however, had their closest relatives in Antarctica, New Zealand and Australia. These species, typically, had a free-swimming larval stage allowing them to disperse short distances (although not necessarily across oceanic depths). Some of these species (like the *Crossaster multispinus* pictured above) fitted a pattern suggested by vicariant distribution, as their relatives are adapted to deep-water living and some are found along the undersea ridge which 'connects' Macquarie Island with Antarctica and New Zealand.

Just four species defied explanation by either dispersal or vicariant spread. These species have relatives only on the southern tip of South America and nearby islands. Perhaps these are the only remnants of ancient Gondwanan fauna that has clung to the edges of the vast continental plates as they have churned their way across the surface of the globe?

12 IS THERE LIFE ON MARS?

A fragment of the meteorite that plummeted to Earth over the small
Victorian town of Murchison on 28 September 1969.
Geology Collection, Museum Victoria.

EXTRAORDINARY CLAIMS DEMAND extraordinary evidence. Take for example, the evidence required to demonstrate paranormal phenomena like telepathy, life after death or telekinesis. In this field, the usual scientific safeguard of replication (whereby contentious findings are not accepted until they have been repeated by other researchers in other laboratories) is not enough. Researchers unwise enough to investigate such a scientifically unpopular concept as telepathy may be submitted to the scrutiny of television entertainers and stage-act magicians for signs of sleight-of-hand and deliberate deceit – an imputation most scientists would find insulting.

A reversal of the rule that extraordinary claims demand extraordinary evidence is that ordinary claims can get away with terribly ordinary evidence. You do not need to call upon sophisticated scientific facts to prove to most people that the sky is blue. The most widely accepted theories require no scientific proof at all. In the fourteenth century little evidence was required to prove that the Earth was flat. In the seventeenth century it was obvious that species were permanent and immutable. In the nineteenth century it was a universal truth that the continents were fixed in their positions. Copernicus's theory of a sun-centred galaxy, Darwin's theory of evolution by natural selection and Wegener's theory of continental drift were all extraordinary theories of their time which demanded (and accumulated) extraordinary evidence.

There is nothing wrong with the natural bias of science towards the conventional and the conservative, but we should always be aware of when and where the bias is operating. When evidence is accepted or rejected on the basis of 'plausibility', we know we are entering the uncharted territories of circumstantial evidence where even lawyers fear to tread. One notable stretch of such uncharted territory can be found in outer space with heated debates over the possibility of life on Mars.

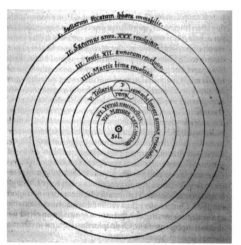

The Copernican galaxy was an extraordinary claim of its times.
History of Science Collections, University of Oklahoma Libraries.

For as long as humans have appreciated the similarities between Earth and its planetary neighbours, the possibility of alien life forms has loomed large in our imagination. In the fourth century BCE, Metrodorus of Chios wrote: 'To consider the Earth as the only populated world in infinite space is as absurd as to assert that in an entire field sown with millet only one grain will grow.' The seeds of life may well be able to grow wherever there is the right environment. And it does seem likely that the 'right' environment may occur elsewhere in the universe at some point in time. But whether or not the seeds of life are likely to appear in these environments is somewhat more contentious.

MARTIAN CANALI

In 1877 Italian scientist Giovanni Schiaparelli (1835–1910) saw a network of lines criss-crossing the Martian surface, which he termed *canali*. Schiaparelli thought these structures were probably natural, but other researchers, particularly the American astronomer Percival Lowell (1855–1916), saw evidence of intelligent life in these canals. Lowell summarised his observations in a book called *Mars* in 1895 (a theme he subsequently expanded upon in other publications and public lectures).

Lowell argued that the conditions on Mars were basically suitable for life, but that there was clearly a lack of surface water. Given such dry but liveable conditions, Lowell felt it could surely be no coincidence that the surface was marked by straight lines linking darkened patches. He argued that the lines were canals fringed by a wide fertile strip of irrigated land, linking cultivated oases visible as darkened nodes. As to the precise nature of the intelligent life form responsible for such structures, Lowell restrained his speculations to a few key observations.

Percival Lowell in his observatory. Lowell Observatory Archives.

Astronomer Christiaan Huygens pioneered the study of the surface of Mars, identifying the planet's polar ice caps.

He noted that the lower gravity of Mars permitted the evolution of much larger creatures than on Earth. Mars was also much older than Earth and may have evolved intelligent life sooner. But the essentially random historical process of evolution made it highly improbable that such life would be humanoid. Even on Earth, Lowell argued, humans were not particularly advanced in evolutionary terms, just fortunate to have evolved a 'mind'. If man had not mastered this cognitive development, 'some lizard or batrachian might just as well have popped into his place'. Lowell concluded that although man will 'probably never find his double anywhere, he is destined to discover any number of cousins scattered through space'.

Of all our neighbouring planets, Mars has caused the most speculation about potential life. Mars has long been regarded as the most Earth-like of this solar system's planets. In 1659 Christiaan Huygens (1629–1695) identified Syrtis Major on the surface of Mars, and thirteen years later he reported the presence of a polar ice-cap similar to those on Earth. At around the same time Giovanni Cassini (1625–1712) calculated that Mars rotated at a similar rate to Earth – providing similar day lengths and seasonal variations. Observations of the Martian surface suggested seasonal changes, while apparently stable features were entered onto maps of the planet's surface (Plate 23). By the late eighteenth century some scientists were reporting that Martian inhabitants would enjoy similar conditions to those found on Earth.

Under the technological constraints of the eighteenth and nineteenth centuries, many of the surface features on Mars were barely discernible. But the inherent pattern-seeking urge of the human psyche drove observers to see and interpret fantastic features on the surface of Mars. Oceans and lakes appeared between continents. Seasonal changes were interpreted as melting ice or shifting vegetational patterns.

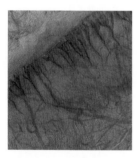

Martian 'canals' have given rise to a great deal of speculation about life on Mars. They are now thought to be a perceptual illusion, created partly by our tendency to interpret random patterns of erosion on the Martian surface as lines.
National Space Science Data Centre, NASA

One popular interpretation of the patterns on the surface of Mars, that they were alien-built canals, was soon discredited by improving technology. The lines on the surface of Mars disappeared at higher resolutions and are now thought to be artefacts of visual perception. At the limits of visibility, dots blur and the human brain 'makes sense' of such random stimuli by grouping the dots into meaningful patterns such as lines. The same perceptual principle sees faces (to which humans are particularly sensitive) emerge from random patterns – whether in craters on the Moon or on shrouds in Turin.

But speculation about the surface of Mars and its capacity for sustaining life continued to circulate until the first images of the Martian surface returned from the Mariner 4 spacecraft in the early 1960s. A succession of Mariner expeditions returned images and information about a barren rocky Martian surface, sub-zero temperatures and a pitifully thin atmosphere comprised primarily of carbon dioxide. With virtually no atmosphere, the sun's ultra-violet rays constantly sterilise the surface, and any water not protected in the polar ice-caps or underground has long ago evaporated. Massive sandstorms periodically lash the Martian landscape with corrosive red dust. Few Earthly life-forms seem capable of surviving such conditions.

At some stage in its past, however, Mars may have been somewhat more hospitable. Huge volcanoes – which make Krakatoa look like an angry pimple – distort the surface while massive canyons and dry riverbeds carve the signature of a watery history. Of course, a planet convulsing with primordial eruptions of molten rock and cataclysmic floods 10,000 times the flow of the Amazon may not suit every organism. But such Hadean conditions do seem to favour some creatures, just as they did in the early history of the Earth. Life on Mars remains a possibility, but has been drastically reduced in scale from Lowell's predictions. If there is life on Mars, it is probably microbial and may well be extinct.

The Syrtis Major region of Mars, reproduced from vol. 14 of the Mars Digital Image Model (MDIM). National Space Science Data Center, NASA.

MODEL ALIENS

Of all the species on Earth, only a handful are capable of surviving the atmospheric conditions on modern Mars. Terrestrial bacteria are even known to survive in space – they drift in the upper limits of our atmosphere far above any other living thing. Bacteria causing the common cold survived a ten-year hiatus in equipment on the Moon before being readily revived and cultured back on Earth.

Bacteria are the oldest life form known on Earth, and their representatives are found in every habitat from ocean depths to deserts to boiling springs in the frozen Arctic. Cyanobacteria identifiable in the fossil record of 3 billion years ago can still be found living in colonial masses or stromatolites in the salty shallow waters of Shark Bay in Western Australia (Plate 24) and in the Gulf of California. But it is their relatives, the archaebacteria, that provide the most plausible models for extraterrestrial life.

Archaebacteria resemble traditional eubacteria, but share similarities with eukaryotes, which developed into multi-cellular organisms. Archaebacteria may even constitute their own kingdom, alongside Monera, Protista, Fungi, Plants and Animals rather than belonging in Monera with the other eubacteria).

Archaebacteria have survived for billions of years under conditions that have previously been regarded as unsuitable for life on Earth. They are survivors from a time when the young planet Earth was very different from today. Archaebacteria are typically anaerobic and are poisoned by oxygen (which is a requisite for most other life forms) Methane-producing archaebacteria, for example, live in anaerobic conditions in swamps and in animals' guts. Salt-loving bacteria populate the inhospitable salt lakes of North America and Australia, their bright red pigmentation providing a sunscreen from the baking rays. Archaebacteria have been found thriving thousands of feet below the sea surface on 'smoking' sulphide chimneys or steam vents. Surviving these conditions is like living in a pressure cooker at 250 degrees C. There is no light, and the increased atmospheric pressure allows water to reach twice boiling point (although these bacteria operate optimally at a temperate 100 degrees C). Archaebacteria can even be found in the freezing waters of Antarctica.

But the most promising model for extra-terrestrial life forms may come from archaebacteria found deep underground – a group that science is just coming to grips with. The presence of subterranean archaebacteria has long been suspected. In the 1920s geologists pondered over the presence of hydrogen sulphide and bicarbonate in water from underground oil fields and proposed that they could only be the product of sulphate-reducing bacteria. But others scoffed at the idea that bacteria could have lived and evolved in oil fields that have been isolated and buried for more than 300 million years.

Deep drilling technology has enabled researchers to reinvestigate subterranean archaebacteria, while minimising the risk of surface contamination of the underground samples. Not only can surface bacteria readily contaminate samples as they are drilled and retrieved, but deep-seated life-forms may also be dependent upon high temperatures and lack of oxygen to survive. As with deep-sea organisms, if such creatures are brought to the surface without maintaining the conditions essential for their survival, the result can be a mangled and uninterpretable mush.

The analysis of pristine rock cores maintained in nitrogen revealed a diverse array of subsurface archaebacteria from rocks as deep as 2.8 kilometres below the surface and living at 75°C. Assuming archaebacteria can live in temperatures up to 110 degrees C (as they do in other environments), they could be found in rocks as deep as 7 kilometres under the Earth's surface. Pockets of bacteria seem to have survived for an extraordinarily long time in isolation from the surface, and they are extraordinarily small. Tiny bacterial rods of *Thermofilum* are just 0.17 micrometres in diameter – thousands could dance on the head of a pin.

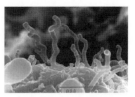

The existence of bacteria the size of the fossil forms found in meteorites is controversial, but recent studies are finding more evidence of smaller and smaller life forms. This electron micrograph image is thought to be of filaments of 'nanobes' or nanobacteria just 100 nanometres in length.
P. Uwins.

Subterranean bacteria on Earth certainly provide a plausible model for Martian bacteria. If any Earth-like life-form is able to survive under Martian conditions it would have to be something like anaerobic bacteria capable of withstanding extreme heat and cold. If micro-organisms once populated the surface of a wetter Mars, they may well have retreated to subterranean water as the surface desiccated.

Such speculation about life on Mars came to a head in August 1996, when American NASA scientists proudly claimed to have 'discovered' life on Mars. No lesser person than President Bill Clinton made their press announcement, claiming the glory of discovering extraterrestrial life as an American first, which, if confirmed, would 'surely be one of the most stunning insights into our universe that science has ever uncovered.'

The meteorite known as ALH84001 is an ancient piece of Martian rock, formed some 4.5 billion years ago. About 3.6 billion years ago, some kind of liquid flowed through the rock, depositing carbonate globules. Sixteen million years ago, ALH84001 was chipped off Mars, probably by a massive meteor strike, and spun into orbit around the Sun. For the next few millennia, the tiny fragment of rock spun through space until eventually, about 13,000 years ago, it collided with Earth and landed in Antarctica. That is where it was found in 1984 by an American meteorite-hunting expedition.

It was not until 1994 that the meteorite was identified as being Martian in origin, with a 'signature' chemical and isotopic make-up of gases trapped in bubbles inside the rock matching those obtained from the Martian atmosphere. The meteorites that periodically shower Earth are typically the product of collisions in space – between comets and planets, moons and more general chunks of space debris – so it is not unusual to find meteorites of Martian origin.

But it was the carbonate globules within the meteorite that excited scientists. Tiny flattened discs of carbonates spread along the walls of cracks and fissures within the meteorite. Although the presence of carbon is not unique to this meteorite, the type of carbon is unusual – polycyclic aromatic hydrocarbons of the kind produced when organisms decay into coal and peat. These Martian hydrocarbons were always found associated with tiny globules in the meteorite that contain traces of iron oxides and iron sulphide, resembling the 'magnetofossils' thought to be produced by terrestrial bacteria. These globules also contain structures which some scientists think might be the remains of tiny fossilised bacteria.

The Martian meteorite ALH 84001 after being collected from Antarctica, with a 1-centimetre cube to show scale.
NASA Johnson Space Center.

The debate over whether or not the traces in the Martian meteorite constituted evidence of bacterial life raged for many months and at a mind-blowing level of technical detail. But one startling fact never seemed to be mentioned. Notwithstanding the President's proud claim, this was not the first time bacteria-like structures had been found in meteorites. Indeed, there have been decades of research on the organic compounds and structures in meteorites, but the papers dealing with ALH84001 never mentioned it. The failure to cite past research on a subject is so unusual in scientific literature that the absence of even a mention of this body of work in the ALH84001 papers is striking.

The history of 'life' in meteorites dates back to the early 1800s, when French chemists discovered significant quantities of carbon in a meteorite which fell in the southern French town of Alais. At the time, carbon was invariably thought to be associated with life, and subsequent papers referred to the material in other carbonaceous meteorites as organic or perhaps even biological in origin. In 1864 another meteorite fell in southern France, in Orgueil, the impact of which was seen and heard across half of the country. The Orgueil meteorite was also found to be 'coal-like'. Coal is produced by the interaction between bacteria and plant material, so to suggest the presence of something resembling coal in a meteorite was highly controversial. Even more surprising was the extraction of hydrocarbons from the Orgueil meteorite. Hydrocarbons are ubiquitous in all life-forms and generally seem to be derived from plant chlorophyll. Since hydrocarbons can be synthesised, however, a non-biological origin cannot be ruled out, and given their abundance and widespread distribution, the presence of hydrocarbons may also be indicative of contamination. Meteorites are often collected after having spent a considerable period of time on the ground, and even after collection they are rarely stored in sterile environments. Museum dust, for example, can contain remarkably high levels of hydrocarbons.

As chemists were continually discovering how to synthesise materials previously thought to emerge only through biological processes, the 'coal-like' nature of the meteorites declined in significance. Scientists generally concluded that most of the carbon compounds in meteorites were probably the result of inorganic processes characteristic of the conditions in an early universe.

Claims of fossilised life-forms in meteorites had been made as early as the 1880s when Otto Hahn (of the same name as the Nobel Prize-winning discoverer of nuclear fission) notoriously claimed to have discovered an entire microcosm of mini-organisms in meteorites ranging from sponges to corals and crinoids. The only problem with this idea was that none of the 'fossils' was more than a millimetre in size. Nonetheless, this discovery received considerable popular attention. Without the original photographs and samples it is impossible to be certain what it was that Hahn actually observed. He may have mistaken minerals for organic structures – yet another example of humans being determined to see meaningful patterns in random events. Or perhaps the meteorites he examined were actually terrestrial rocks containing genuine remains of life. At the time, the small size of the so-called fossils seemed to rule out the possibility that they were genuine. Interest in the meteorites as potential harbingers of extraterrestrial life was high at the time, with even microbiological pioneer Louis Pasteur (1822–1895) examining the Orgueil meteorite for bacterial traces (without success).

For many years, debate raged about whether the organic compounds found in these meteorites, including hydrocarbons, fatty acids, amino acids, porphyrins and nucleotide bases, were produced in space by unknown non-biological processes. Or were they due to contamination of the meteorite by terrestrial sources? Or were they both extraterrestrial and biological in origin? Many scientists felt that their work suggested neither abiotic nor contaminative sources, but few felt they had enough evidence to suggest extraterrestrial life. Only the very brave, like Frederick Hoyle and Chandra Wickramasinghe, were willing to suggest that life may have originated, not on Earth, but in space (a theory most comprehensively summarised in *Space Travellers* but most succinctly expressed in *The Origin of Life*).

It was not until the 1960s that research into meteorite fossils really took off. Scientists from a wide variety of disciplines described and debated the nature of the tiny oval and tubular structures found in many meteorites. The more alien of these structures were argued to be artefacts of mineralisation, while the ones which looked more convincingly biological were explained away by contamination. Pollen grains, even furnace ash, were argued to be the source of contamination in some cases.

Even when many individual cases of 'microfossils' could be explained away by common terrestrial processes, some scientists were not yet prepared to shut the door on the possibility of extraterrestrial life. This lively period of research was carefully documented in Bartholomew Nagy's classic tome *Carbonaceous Meteorites*. In the foreword to this book Professor Harold Urey suggests, 'Perhaps it is well not to draw conclusions, but it appears most likely that the carbonaceous compounds and small carbonaceous objects found in these meteorites would be confidently assumed to be of biological origin if found on Earth.' The problem was not so much with *what* was found, but where it was found – circumstance, not evidence, dictated the outcome.

Illustration of a 'fossil bacteria'.
B. Nagy (1975).

MURCHISON METEORITE

On a fine spring morning, 28 September 1969, at a few minutes to 11 o'clock, a new meteorite plummeted into the debate. A large orange fireball, trailed by a smoking tail and loud crackling boom, swept across the northern farming district of Victoria and splattered a shower of small rocks across paddocks and buildings around the small town of Murchison. The significance of these scattered bits of rock was appreciated immediately, and many locals searched diligently for remains of the 'Murchison meteorite'. This rapid collection of specimens after impact reduced the potential for terrestrial contamination with soil bacteria.

Many of these early collectors reported a strange smell, like methylated spirits, which scientists later identified as the organic compound pyridine – the substance added to methylated spirits in order to make it unpalatable to drink. The Murchison meteorite was soon identified as one of the largest carbonaceous meteorites of its kind – its pre-entry weight was estimated at 2 tonnes, although the largest fragment recovered weighed a mere 189 grams. Fragments of the meteorite were ultimately distributed to scientific institutions around the world.

Hans-Dietrich Pflug from the Justus-Liebig University in Germany continued the early work on 'microfossils' using the Murchison meteorite, the large carbonaceous meteorite collected in Victoria. Pflug examined demineralised slices of meteorite using an electron microscope and provided further suggestions that the tubular and spherical structures in meteorites may well be biological in origin. By 1980, Ed Roedder demonstrated the presence of carbonates associated with these structures.

Over the past decade or two, these microstructures have increasingly been compared to microstructures found in the oldest rocks on Earth from Greenland and Western Australia (4000 million years old). Similar structures are found in later Precambrian rocks associated with iron bacteria. Elements in early Precambrian rocks like the South African Onverwacht sediments (3400 million years old) have been interpreted as fossil algae or fossil bacteria. The evidence for a biological origin of structures in later rocks like those from Bitter Springs in central Australia (1000 million years old) is even stronger, and few dispute that these represent the earliest traces of life. Yet there is little to distinguish these 'bacterial' and 'algal' fossils from the structures found in meteorites. If the meteorite structures were found in terrestrial rocks – in the right place, at the right time – they would probably be assumed to be bacterial remains.

The major factor holding back acceptance that these meteorites contain possible life traces is not in the traces themselves, but in the nature of the meteorites. Most of the carbonaceous meteorites have no obvious planetary origins, but are thought to be the remains of spent comets and asteroids. The Murchison meteorite, for example, was only a meteor for about 800,000 years, but dates its origins to the asteroid belt over 4500 million years ago – as old as the beginning of the universe. Scientists might be prepared to speculate that life might occur on other planets, but few are prepared to consider the possibility that life could survive, much less evolve, within the inhospitable asteroid belt. The possibility of life traces in meteorites of a non-planetary origin is therefore ruled out on circumstantial grounds – they have been in the wrong place at the wrong time.

But the extraordinary evidence continues to grow. In 2001, a pristine, uncontaminated piece of the Orgueil meteorite was re-examined for trace levels of amino acids – the building blocks of DNA. The researchers found the two amino acids present were not Earthly in origin but were almost certainly synthesised in space and probably in a comet. They also examined the Murchison meteorite (among others), finding that it contained an even more complex mix of more than seventy types of amino acids, probably with an asteroid origin. NASA's recent Deep Impact mission, in which a projectile deliberately collided with the comet Tempel 1 and penetrated about 10 metres into its surface, revealed an abundance of organic compounds. Increasingly, it is becoming apparent that the organic building blocks of life are littered across the cosmos. But what of life itself?

In 2004 more evidence emerged of extraterrestrial life when Richard Hoover re-examined the interior of the Orgueil meteorite and photographed the lithified and carbonised structures. These, he argues, are consistent with mats of cyanobacteria such as *Phormidium tenuissimum* and sulphur bacteria, which must therefore have grown on the meteorite's parent body before its arrival on Earth.

As our exploration of other planets, comets and asteroids continues, further suggestions of biological processes in space continue to amass, slowly but surely shifting the leviathan of scientific opinion towards the idea that bacterial life may well occur elsewhere in space. Whatever is found on future Mars or space missions, and whatever is concluded from these findings, we should guard against the tendency to accept evidence that conforms to our preconceptions of life in space and reject similar evidence for theories which seems too far-fetched to be reasonable. The fossil evidence for life on Mars is as good as the early fossil evidence for life on Earth and as good as the evidence for life in asteroids. Circumstantial evidence seems to be the only way to decide which traces are genuine. The circumstantial evidence of life on Earth is so strong as to be irrefutable (we are; therefore it was). But I am wary of circumstantial evidence based on our knowledge of what conditions occur on other planets or in outer space – we simply do not know much about it, and every space mission changes our assumptions.

Even though studying bacteria should be much easier than space research, I am even more cautious about placing limitations on bacterial adaptability based on terrestrial research. Bacteria are far and away the most successful kingdom of creatures on Earth – they have lived the longest, are the most adaptable, occupy the greatest extremes of environmental conditions, and are the most diverse and abundant. And we know absolutely nothing about the vast majority of them. Who knows what environments they (or similar organisms) could adapt to elsewhere?

Metrodorus might have been right in suggesting we are just one seed of many sown in a universal field. But even if the precursors to life are scattered throughout the cosmos, it does not necessarily make it any more likely that complex multicellular life-forms have evolved elsewhere. Bacteria existed on Earth for at least 3 billion years before the evolution of multicellular life-forms. Only once life had harnessed the structural capacities of calcium to make skeletons did the explosion occur into the diversity of life forms and structures that we know today. But the key to what precipitated this Cambrian explosion is yet another of life's great mysteries; without understanding that, we cannot calculate the odds of finding complex multicellular life forms on other planets. Percival Lowell was probably right that we will never find our double, but it may take a lot longer than he expected to find any relatives more similar to us than a bacterium.

SOURCES

1. CURIOUS COLLECTIONS

The quote 'nothing in biology makes sense except in the light of evolution' is from T. Dobzhansky (1973), Nothing in biology makes sense except in the light of evolution, *American Biology Teacher*, 35: 125–9. The role of museums and natural history collections in the development of the biological sciences is discussed further in K. R. Arnold (1992), Cabinets for the curious: practising science in early modern English museums, PhD thesis, Princeton University; T. Bennet (2004), *Pasts beyond memory: evolution, museums and colonialism*, Routledge, London; R. H. Belk (1995), *Collecting in a consumer society*, Routledge, New York; and E. Hooper-Greenhill (1992), *Museums and the shaping of knowledge*, Routledge, London. The role of Australian and colonial museums is discussed in T. Bonyhady (2000), *The colonial earth*, Miegunyah Press, Melbourne; S. G. Kohlstedt (1983), Australian museums of natural history: public priorities and scientific initiatives in the nineteenth century, *Historical Records of Australian Science*, 5 (4): 1–29; and S. Sheets-Pyenson (1988), *Cathedrals of science: the development of colonial natural history museums during the late nineteenth century*, McGill–Queens University Press, Montreal. A history of the Melbourne Museum can be found in R. T. M. Pescott (1954), *Collections of a century: the history of the first hundred years of the National Museum of Victoria*, National Museum of Victoria, Melbourne; and C. Rasmussen (2001), *A museum for the people: a history of Museum Victoria and its predecessors, 1854–2000*, Scribe Publications, Melbourne.

2. A BEAST NAMED SU

Joseph Banks' 'an animal … very swift' and 'nothing certainly … resembles him' is from R. Parkin (1997), *H.M. Bark Endeavour: her place in Australian history*, Melbourne University Press, Carlton, pp. 336, 339. James Cook's 'I could have … like a hare or deer' is from the same book, p. 337. Eighteenth-century English reactions to kangaroos are discussed in R. M. Younger (1988), *Kangaroo: images through the ages*, Hutchinson Australia, Melbourne; and B. Smith and A. Wheeler (eds) (1988), *The art of the First Fleet and other early Australian drawing*s, Oxford University Press, Melbourne. George Bennett's response to Charles Lamb's reputed comment on kangaroos is from G. Bennett (1860), *Gatherings of a naturalist in Australasia*, John van Voorst, London, p. 5.

For further information on the scientific debate surrounding the name of the Grey Kangaroo see J. H. Calaby, G. Mack and W. D. L. Ride (1962), The application of the generic name Macropus Shaw 1790 and of other names commonly referred to the Grey Kangaroo, *Memoirs of the Queensland Museum* 14 (2): 25–31; see also H. J. Frith and J. H. Calaby (1969), *Kangaroos*, F. W. Cheshire, Melbourne. Information on the Aboriginal origin of the name 'kanguru' is from R. M. W. Dixon, W. S. Ramson and M. Thomas (1990), *Australian Aboriginal words in English: their origin and meaning*, Oxford University Press, Melbourne, pp. 67–8.

For a general account of the contributions of European explorers until Cook on Australian biology (including de Jode and Pinzó), see G. P. Whitley (1970), *Early history of Australian zoology*, Royal Zoological Society of New South Wales, Sydney. Cornelis de Bruijn published an account of his travels in C. de Bruijn (1720), *Voyage to the Levant and travels into Muscovy, Persia, and the East Indies*. 'The Filander … neck thrust out of this bag' is from R. M. Younger (1988), *Kangaroo: images through the ages*, Hutchinson Australia, Melbourne, pp. 42–3. William Dampier's quote 'with short forelimbs' is from A. S. George (1999), *William Dampier in New Holland: Australia's first natural historian*, Bloomings Books, Melbourne, p. 131. Quotes from the de Vlamingh expedition are all from P. Playford (1998), *Voyage of discovery to Terra Australis, by William de Vlamingh 1696–97*, Western Australian Museum, Perth, as follows: 'rats nearly as big as cats' with 'a pouch below their throats … to what end nature had created the animal like this', Witson, pp. 28–9; 'had great pleasure … pleasurable beyond all islands I have ever seen' and 'a kind of bag or purse hanging from the throat upon the breast downwards', de Vlamingh, p. 29; and 'as a signal of farewell to the miserable South Land', Torst, p. 62. Volkerson (or Volkertzoon), 'a civet-cat, but with browner hair' is quoted in G. P. Whitley (1970), *Early history of Australian zoology*, Royal Zoological Society of New South Wales, Sydney, p. 62. Pelsaert's account of his voyage is in F. Pelsaert (1647), *The voyage of the Batavia*, trans. W. Siebenhaar (1897; republished 1994), Hordern House, Sydney. The Pelsaert quote 'on these islands … when they are hunted' is from H. Drake-Brockman (1963),

Voyage to disaster: the life of Francisco Pelsaert, Angus & Robertson, Sydney, pp. 235–6.

A discussion of the earliest representations of opossums (including Pinzó, Waldseemüller, Gesner and Topsell) can be found in C. Hartman (1952), *Possums*, University of Texas, Austin. Tyson's dissection of an opossum is from E. Tyson (1698), *The anatomy of an opossum: dissected at Gresham College*, Royal Society, London. For further information on the Dieppe maps see H. Wallis (1988), Java la Grande: The enigma of the Dieppe Maps, in G. Williams and A. Frost (eds), *Terra Australis to Australia*, Oxford University Press, Melbourne, pp. 39–81. Portuguese activity in Australia is discussed in K. G. McIntyre (1982), *The secret discovery of Australia: Portuguese ventures 250 years before Captain Cook*, Picador; Sydney (rev. and abridged edn), although there have been numerous criticisms of this work. The account of a cuscus, 'Some animals resemble ferrets … people eat them like rabbits, seasoned with spices' is from H. T. T. M. Jacobs, SJ (1971), *A treatise on the Moluccas (c. 1544)*, Jesuit Historical Institute, Rome, p. 61.

For information on the distribution of marsupials in the south-west Pacific and Moluccan Islands see T. F. Flannery (1995), *Mammals of the South-West Pacific and Moluccan Islands*, Reed Books, Chatswood, NSW. The standard reference for Australian mammals is R. Strahan (1990), *The mammals of Australia*, Reed Books, Chatswood, NSW.

3. LOCAL KNOWLEDGE

The reference to Isaac Newton relates to his statement 'If I have seen further it is by standing on ye shoulders of Giants', in a letter to Robert Hooke on 5 February 1675, although there is some doubt as to whether he was being modest or commenting on Hooke's lack of stature. For further information on Elizabeth Gould see C. Jordan (2005), *Picturesque pursuits: colonial women artists and the amateur tradition*, Melbourne University Press, Melbourne. On Amalie Dietrich see R. Sumner (1993), *A woman in the wilderness: the story of Amalie Dietrich in Australia*, New South Wales University Press, Sydney; on Edith Coleman see D. Clode (2005), Popular and professional communicators: Edith Coleman and Norman Wakefield, *Victorian Naturalist*; on Jeanne Baret see J. Dunmore (2002), *Jeanne Baret, first woman around the world (1766–1768)*, Auckland, Heritage Press; and on Jane Franklin see A. Selzer, (2002), *Governors' wives in colonial Australia*, National Library of Australia, Canberra. The George Eliot quote is from G. Eliot (1871), *Middlemarch*, republished 1985, Penguin Classics, Aylesbury, p. 896.

Further information on George Bass's exploits as an explorer and his interactions with indigenous people can be found in K. M. Bowden, (1952), *George Bass, 1771–1803: his discoveries, romantic life and tragic disappearance*, Oxford University Press, Melbourne. Blandowski's expedition is described in W. Blandowski (1857), Recent discoveries in Natural History on the Lower Murray, *Philosophical Institute of Victoria*, 2 (1): 124–37; see also D. Clode (2001), The respectful invader, in C. Rasmussen, *A museum for the people: A history of Museum Victoria and its predecessors, 1854–2000*, Scribe Publications, Melbourne, pp. 18–20; and P. Stanbury (1975), *100 years of Australian scientific explorations*, Holt, Rinehart & Winston, Sydney.

Information on Arfak highlanders is derived from a 1928 study of New Guinean birds by Ernst Mayr, quoted in E. O. Wilson (1992), *The diversity of life*, Penguin, London, pp. 38–9. Material on Groote Eylandt classifications comes from J. A. Waddy (1988), *Classification of plants and animals from a Groote Eylandt Aboriginal point of view*, North Australia Research Unit, Australian National University, Canberra. Howitt's species lists are from I. Mansergh and L. A. Hercus (1981), An Aboriginal vocabulary of the fauna of Gippsland, *Memoirs of the National Museum Victoria*, 42: 107–22. Other indigenous word lists can be found in J. Dixon and L. Huxley (1985), *Donald Thomson's mammals and fishes of Northern Australia*, Nelson, Melbourne; R. M. W. Dixon, W. S. Ramson and M. Thomas (1990), *Australian Aboriginal words in English: their origin and meaning*, Oxford University Press, Melbourne. Information on the Adnymathanha is derived from D. Tunbridge, (1991), *The story of the Flinders Ranges mammals*, Kangaroo Press, Kenthurst, NSW.

Information from Spencer's expedition is from W. B. Spencer (1896) (ed.), *Report on the work of the Horn Scientific Expedition to Central Australia: Part 1.—Introduction, Narrative and summary of results, supplement to zoological report, map*, Melville, Mullen & Slade, Melbourne; and W. B. Spencer (1896) (ed.), *Report on the work of the Horn Scientific Expedition to Central Australia: Part 2.—Zoology*. Melville, Mullen & Slade, Melbourne, including the quote 'While the Urgătta...pulling it out with the hand.' Additional information on central Australian vertebrates is from J. R. W. Reid, J. A. Kerle and S. R. Morton (1993), Uluru fauna: the distribution and abundance of vertebrate fauna of Uluru (Ayers Rock–Mount Olga) National Park, NT, *Kowari*, 4. The results of the new Horn expedition are documented in A. L. Yen, J. Gillen, R. Gillespie, R. Vanderwal, and the Mutitjulu Community (1997), A preliminary assessment of Anangu knowledge of central Australian invertebrates, *Memoirs of the Museum of Victoria*, 546(2): 631–4.

Information on desert lizards is from E. R. Pianka (1986), *Ecology and natural history of desert lizards: analyses of the ecological niche and community structure*, Princeton University Press, Princeton.

4. WATER, WATER EVERYWHERE

An example of Sydney's panic over its water can be seen on the front page of *The Sydney Morning Herald*, 30 July 1998, headlined 'Most of Sydney told: boil drinking water'. The official reaction to the crisis can be found in P. McClellan (1998), *Sydney Water Inquiry: final report*, NSW Premier's Department, Sydney; and NSW Department of Health (1998), The Sydney water incident: July–September 1998, *NSW Public Health Bulletin* 9: 91–4. The value of closed catchments and the example of the Catskill Mountain dilemma is from K. Ellison (2003), Investing in nature, *The Source* (Melbourne Water), 26.

For a history of Melbourne's water board see T. Dingle and C. Rasmussen (1991), *Vital connections: Melbourne and its Board of Works, 1891–1991*, McPhee Gribble, Melbourne. Accounts of Melbourne's early history are derived largely from M. Cannon (1991), *Old Melbourne Town: before the gold rush*, Loch Haven, Melbourne.

Richard Marchant's research on the effect of dams on invertebrate river life can be found in R. Marchant (1987), *Thomson River post-construction monitoring program: changes in the benthic invertebrate communities of the Thomson River, Victoria, after dam construction*, A report to the Department of Water Resources, Victoria; R. Marchant (1988), Changes in the benthic invertebrate communities of the Thomson River, Southeastern Australia, after dam construction, *Regulated Rivers: Research and Management*, 4: 71–89; and R. Marchant and G. Hehir (2002), The use of AUSRIVAS predictive models to assess the response of lotic macroinvertebrates to dams in south-east Australia, *Freshwater Biology*, 47: 1003–50.

Debates over the role of forests in water catchment are sourced from M. V. Pollio (27 BCE), *On architecture*, Book VIII, ch. 1, trans. J. Gwilt (1826), *The Architecture of Marcus Vitruvius Pollio*, Priestley & Weale, London. Ferdinand von Mueller's argument 'a heritage ...carefully maintained' is from F. Mueller (1871), *Forest culture in relation to industrial pursuits*, Government Printer, Melbourne, p. 96. Modern research on water loss and retention by forest is from G. M. Dunn and D. J. Connor (1991), *Management of transpiration loss and water yield in Mountain Ash: a final report*, Faculty of Agriculture and Forestry, University of Melbourne, Melbourne; and R. Lee (1980), *Forest hydrology*, Columbia University Press, New York.

The results of the Port Phillip Bay survey are summarised in Commonwealth Scientific and Industrial Research Organisation (1996), *Port Phillip Bay environmental study: final report*, CSIRO, Canberra.

5. FORESTS OF FIRE

Dumont D'Urville's account of his expedition is in J. Dumont D'Urville (1830–5), *Voyage de la corvette l'Astrolabe, exécuté par ordre du Roi pendant les années 1826, 1827, 1828, 1829, sous le commandement de M. Jules S-C Dumont D'Urville*, 17 vols, Tastu & Cie, Paris. His quote 'a lovely grassland … rather like our royal forest around Paris' is from the translation by H. Rosenman (1987), *Two voyages to the South Seas by Captain Jules S-C Dumont D'Urville*, Melbourne University Press, Melbourne, vol. I, p. 56. A summary of research on landscape preferences can be found in D. Clode and M. Burgman (1997) (eds), Cultural values of old-growth forests, in *Joint old-growth forest project: summary report*, NSW Parks and Wildlife Service and NSW State Forests, Sydney.

An overview of the history of the Mountain Ash forests can be found in T. Griffiths (2001), *Forests of ash: An environmental history*, Cambridge University Press, Melbourne. For further information on the biodiversity of Mountain Ash forests see L. D. Ahern and A. L. Yen (1977), A comparison of the invertebrate fauna under Eucalyptus and Pinus forests in the Otway Ranges, Victoria, *Proceedings of the Royal Society of Victoria*, 89: 127–36; G. W. Brown et al. (1989), *Flora and fauna of the Acheron forest block, Central Highlands, Victoria*, Department of Conservation, Forests and Lands, Melbourne; and T. M. Howard (1975), Litter fauna in *Nothofagus cunninghamii* forests, *Proceedings of the Royal Society of Victoria*, 87: 207–13.

The importance of disturbance in Australian forests is summarised in P. M. Attiwill (1994), The disturbance of forest ecosystems: the ecological basis for conservation management, *Forest Ecology and Management*, 63: 247–300; and P. M. Attiwill (1994), Ecological disturbance and the conservative management of eucalypt forests in Australia, *Forest Ecology and Management*, 63: 301–46. Further information on the role of fire in the Australian landscape can be found in S. J. Pyne (1991), Antipodal fire: bushfire research in Australia and America, in R. W. Home and S. G. Kohlstedt (eds), *International science and national scientific identity*, Kluwer, Boston, pp. 225–87.

Darwin's quote comes from C. Darwin (1845), *Journal of researches into the natural history and geology of the countries visited during the voyage of the HMS 'Beagle' round the world*, Ward, Lock & Co., London, p. 409. The Lawson quote is from H. Lawson (1896), His country after all, *While the billy boils*, Angus & Robertson, Sydney.

The case study of lizards is derived from information in G. W. Brown and J. L. Nelson (1992), *Habitat utilisation by heliothermic reptiles of different successional stages of* Eucalyptus regnans *(Mountain Ash) forest in the central highlands,* VSP Technical Report No. 17, Department of Conservation and Natural Resources, Melbourne. Information on Leadbeater's Possum is from D. Lindenmayer (1996), *Wildlife and woodchips: Leadbeater's Possum, a test case for sustainable forestry*, University of NSW Press, Sydney.

6. THE REAPPEARING POSSUMS

A history of introduced species can be found in E. Rolls (1984), *They all ran wild: the animals and plants that plague Australia*, Angus & Robertson, Sydney. Basic biological details on Australian mammals can be found in R. Strahan (1995), *Mammals of Australia*, Australian Museum, Sydney.

Specific details on the original description of Leadbeater's Possum can be found in R. Broom (1895), On a small fossil Petaurus-like Marsupial, *Proceedings of the Linnaean Society of New South Wales*, series 2, 10(4): 568–70; W. D. L. Ride (1970), *A guide to native mammals of Australia*, Oxford University Press, Oxford; and F. McCoy (1867), On a new genus of phalanger, *Annals of the Magazine of Natural History*, 3: 287–8. The subsequent searches for more specimens are documented in W. B. Spencer (1899), Remarks on a rare marsupial, *Victorian Naturalist*, 16 (7): 105–6; C. W. Brazenor (1932), A re-examination of *Gymnobelideus leadbeateri* McCoy, *Australian Zoologist*, 7 (2): 106–9; and C. W. Brazenor (1946), Last chapter to come: A history of Victoria's rarest possum, *Wild Life*, 8, 383–4. Brazenor's quote that the Leadbeater would 'have none of ....tops of larger trees' is from the latter article, p. 383.

The rediscovery of living Leadbeater's Possums was documented in H. E. Wilkinson (1961), The rediscovery of Leadbeater's Possum, *Gymnobelideus leadbeateri* McCoy, *Victorian Naturalist*, 78: 97–102. Modern biological information on Leadbeater's Possum can be found in D. Lindenmayer (1996), *Wildlife and woodchips: Leadbeater's Possum, a test case for sustainable forestry*, University of NSW Press, Sydney; and A. P. Smith (1980), The diet and ecology of Leadbeater's Possum and the Sugar Glider, PhD thesis, Monash University, Melbourne. D. Clode (2001), The elusive Leadbeater, in C. Rasmussen, *A museum for the people: a history of Museum Victoria and its predecessors, 1854–2000*, Scribe Publications, Melbourne, pp. 253–5, provides a summary of the story.

Details on the original discovery of Mountain Pygmy Possum fossil remains can be found in R. Broom (1895), On a small fossil marsupial with large grooved premolars, *Proceedings of the Linnaean Society of New*

South Wales, series 2, 10(4): 563–7. W. D. L. Ride (1956), The affinities of *Burramys parvus* Broom, a fossil marsupial, *Proceedings of the Zoological Society of London*, 127 (3): 413–29, provides details of the species affinities. Biographical information on Norman Wakefield is from D. Clode (2002), Norman Wakefield, in J. Ritchie and D. Langmore (eds), *Australian Dictionary of Biography*, vol. 16, Melbourne University Press, Melbourne.

Further details on the rediscovery of living Mountain Pygmy Possums can be found in J. Calaby, H. Dimpel and Cowan I. McTaggart (1971), *The Mountain Pigmy Possum* Burramys parvus *Broom (Marsupialia) in the Kosciusko National Park, New South Wales*, Division of Wildlife Research Technical Paper No. 23, CSIRO; and J. M. Dixon (1971), *Burramys parvus* Broom (Marsupialia) from Falls Creek area of the Bogong High Plains, Victoria, *Victorian Naturalist*, 88: 133–8. Moden biological information on the species is derived from I. Mansergh and L. Broome (1994), *The Mountain Pygmy Possum of the Australian Alps*, New South Wales University Press, Sydney. Data on the species genetics is from M. J. Norman, J. A. Osborne, L. Christidis and N. D. Murray (2000), Genetic distinctness of isolated populations of an endangered marsupial, the Mountain Pygmy-possum, *Burramys parvus*, *Molecular Ecology*, 9 (5): 609–13. For information on the Mahogany Glider see S. van Dyck (1993), The taxonomy and distribution of *Petaurus gracilis* (Marsupialia: Petauridae), with notes on its ecology and conservation status, *Memoirs of the Queensland Museum* 33: 77–122.

7. THE MISSING MOLLUSC

I was first alerted to the interesting story of the trigonia in W. J. Dakin (1976), *Australian Seashores*, Angus & Robertson, Sydney, pp. 293–4, which eventually led me to S. J. Gould (1993), A foot soldier for evolution, *Eight little piggies: reflections in natural history*, Norton, New York, pp. 439–55. The Gould quote 'must have known … publish his description' is from p. 451. Basic information on the trigoniidae is from T. A. Darragh (1986), The Cainozoic trigoniidae of Australia, *Alcheringa*, 10: 1–34; and T. S. Hall (1901), Growth stages in modern Trigonias belonging to the section Pectinatae, *Proceedings of the Royal Society of Victoria*, 14: 17–21.

François Péron's quote is from F. Péron (1809), *A voyage of discovery to the southern hemisphere, performed by order of the Emperor Napoleon during the years 1801, 1802, 1803 and 1804*, vol. 1, Richard Phillips, London, republished 1975, Marsh Walsh, Melbourne, pp. 187–8. Original descriptions of the collecting of the trigonia by Quoy and Gaimard are from J. Dumont d'Urville (1830–35), *Voyage de la corvette l'Astrolabe, exécuté par ordre du Roi pendant les années 1826, 1827, 1828, 1829, sous le commandement de M. Jules S-C Dumont d'Urville*, 17 vols, Tastu & Cie, Paris. The quote that they 'left Westernport ... looking for' is translated from vol. 1: *Histoire*, p. 144. Dumont d'Urville's quotes, 'the drag-net ... for some time' and 'their most necessary belongings', are from H. Rosenman (1992), *Two voyages to the South Seas: Captain Jules S-C Dumont d'Urville*, Melbourne University Press, Melbourne, vol. I, pp. 65, 112. Rosenman's translations provide an accessible account of the narrative of the journey, but the original must be consulted for the scientific accounts. Quoy's comment 'We were so anxious ... from our collection' is from S. J. Gould (1993), A foot soldier for evolution, *Eight little piggies: reflections in natural history*, Norton, New York, p. 442.

Early scientific discussions of trigonia's significance to various evolutionary and creationist theories include J. P. B. Lamarck (1804), Sur une nouvelle espèce de Trigonie et sur une nouvelle espèce d'Huître, découverte dans le voyage du capitaine Baudin, *Annales du Muséum d'Histoire Naturelle*, 4: 353–4. The quote 'small species ... are truly extinct' is from J. P. B. Lamarck (1809), *Philosophie zoologique*, Dentu, Paris, p. 45. Darwin's view that 'A single species of Trigonia ... than its production' is from C. Darwin (1859), *On the origin of species by means of natural selection*, John Murray, London, p. 321. L. Agassiz (1840), Mémoire sur les trigonies, *Etudes critiques sur les mollusques fossiles*, Petitpierre, Neuchatel, is quoted 'The absence of Trigonia ... from each other' in S. J. Gould (1993), *Eight little piggies: reflections in natural history*, Norton, New York, p. 448.

Background information on Australian scientists at the time was derived from A. Mozely-Moyal (1976), *Scientists in nineteenth-century Australia: a documentary history*, Cassell Australia, Melbourne. Lyell's account of Lamarck's transmutationist theory is from C. Lyell (1830), *Principles of geology*, 3 vols, London.

The discovery of the missing Cainozoic trigonia is documented in H. M. Jenkins (1865), On the occurrence of a Tertiary species of Trigonia in Australia, *Quarterly Journal of Science*, 2 (April): 362–4; and H. M. Jenkins (1866), On the occurrence of a recent species of Trigonia (*T. lamarckii*) in Tertiary deposits in Australia, *Geological Magazine*, 3: 201–3. Quotes reflecting McCoy's opposition to evolutionary ideas can be found in F. McCoy (1870), *Lecture on the order and plan of creation, delivered before the Early Closing Association, Melbourne 1869–70*, pp 23, 25, 26–7, 31–2, National Library of Australia. McCoy's 'furnished an extraordinary ... still more recent epoch' and 'urged to make ... Mr. Jenkins' paper' are from F. McCoy (1875), On a third new Tertiary species of Trigonia, *Annals and Magazine of Natural History*, ser. 4, 15: 316–17.

8. BRAINBOX

A general introduction to Australia's polar dinosaurs can be found in P. Vickers-Rich and T. H. Rich (1993), Australia's polar dinosaurs, *Scientific American*, July: 40–5. Further information is derived from T. H. Rich, P. Vickers-Rich and R. A. Gangloff (2002), Polar dinosaurs, *Science*, 295: 979–80; P. V. Rich et al. (1988), Evidence for low temperatures and biologic diversity in Cretaceous high latitudes of Australia, *Science*, 242: 1403–6; T. H. Rich and P. V. Rich (1989), Polar dinosaurs and biotas of the early cretaceous of southeastern Australia, *National Geographic Research*, 5 (1): 15–53; E. Buffetaut (2004), Polar dinosaurs and the question of dinosaur extinction: a brief review, *Palaeogeography, Palaeoclimatology, Palaeoecology*, 214 (3): 225–31; and A. Constantine et al. (1998), Periglacial environments and polar dinosaurs, *South African Journal of Science*, 94: 137–41.

Information on the neurology of brains, particularly the optic tectum and pineal system, has been sourced from R. A. Barton, A. Purvis and P. H. Harvey (1995), Evolutionary radiation of visual and olfactory brain systems in primates, bats and insectivores, *Philosophical Transactions of the Royal Society of London B*, 348: 381–92; A. B. Butler and W. Hodds (1996), *Comparative vertebrate neuroanatomy: Evolution and adaptation*, Wiley–Liss, New York; H. D. Potter (1969), Structural characteristics of cell and fiber populations in the optic tectum of the frog (*Rana catesbeiana*), *Journal of Comparative Neurology* 136: 203–32; P. S. Ulinski, D. M. Dacey and M. I. Sereno (1992), Optic tectum, in C. Gans and P. S. Ulinski (eds), *Biology of the Reptilia*, vol. 17: *Neurology C*: 241–366; and R. Huber et al. (1997), Microhabitat use, trophic patterns and the evolution of brain structure in African cichlids, *Brain Behavior and Evolution*, 50 (3): 167–82. The quote 'most rightly …. pre-Tertiary periods' is from W. B. Spencer (1885), *On the presence and structure of the pineal eye in Lacertilia*, J. E. Adlard, London, p. 27.

Information on hypsilophodontids and dinosaurs generally has been derived from P. M. Galton (1974), The Ornithiscian dinosaur Hypsilophodon from the Wealden of the Isle of Wight, *Bulletin of the British Museum (Natural History) Geology*, 25: 1–149; J. A. Hopson (1977), Relative brain size and behavior in archosaurian reptiles, *Annual Review of Ecology and Systematics*, 8: 429–48; T. H. Huxley (1869), On Hypsilophodont, a new genus of Dinosauria, *Abstracts of the Proceedings of the Geological Society of London*, 204: 3–4; and J. J. Roth and E. C. Roth (1980), The parietal-pineal complex among paleovertebrates: evidence for temperature regulation, in R. D. K. Thomas and E. C. Olson (eds), *A cold look at warm-blooded dinosaurs*, West View Press, British Columbia, pp. 189–231.

9. THE APE CASE

Charles Darwin's work has been published in C. Darwin (1859), *On the origin of species by means of natural selection*, John Murray, London; and C. Darwin (1875), *The descent of man*, John Murray, London. For differences between Darwin and Wallace on natural selection see A. R. Wallace (1857), On the law which has regulated the introduction of new species, *Annals and Magazine of Natural History*, series 2, 14: 184–96. For Huxley's views on the implications of Darwin's theory for humans see T. H. Huxley (1863), *Evidence as to man's place in nature*, Williams & Norgate, London, with the quote 'Perhaps no order ... he is but dust' being from p. 98. Tyson's anatomical work on apes is published in E. Tyson (1699), *Orang-outang, sive Homo Sylvestris: or, the Anatomy of a pygmie*, facs. edn 1966, Dawsons of Pall Mall, London, including the quote 'our Pygmie ... that I know of'. For an account of the debate between Huxley and Wilberforce, see J. R. Lucas (1979), Wilberforce and Huxley: a legendary encounter, *The Historical Journal*, 22 (2): 313–30, with 'rather have a miserable ape ... a grave scientific discussion' (from Huxley's letter to Dyster, 9 September 1860) on p. 21. Richard Owen's views on Darwin and Huxley are reported in many of his publications, but for an example of his comparative work with ape and human brains see R. Owen (1861), The gorilla and the negro, *The Athenaeum*, 23 March: 395–6. The quote by Herbert Spencer 'progress ....part of nature' is from H. Spencer (1851), *Social statics, or, The conditions essential to human happiness specified, and the first of them developed* John Chapman, London, vol. VI, ch. 2, p. 4.

For a further discussion of the Australian argument over human evolution, see B. W. Butcher (1988), Gorilla warfare in Melbourne: Halford, Huxley and man's place in nature, in R. W. Home (ed.), *Australian science in the making*, Cambridge University Press, Cambridge, pp. 153–69. Halford's arguments are published in G. B. Halford (1863), *Not like man bimanous and biped, nor yet quadromanous, but cheiropodous*, Wilson & Mackinnon, Melbourne (whence the quote 'Surely the intricacies .... one from the other', p. 15); and G. B. Halford (1864), *Lines of demarcation between man, gorilla and macaque*, Wilson & Mackinnon, Melbourne. McCoy's support for Halford and his position are evidenced by F. McCoy (1863), Letter to Prof. Halford, 13 July 1863, National Museum Outward Letter Book, vol. 2, Sept 1861–May 1865, p. 217; and F. McCoy (1865), Gorillas at the National Museum, *Argus*, 20 June: 5, which contains the quote 'how infinitely remote ... quadrumana and man'.

The quotes 'Whatever existing species ... other existing species' and 'that of Apes ... and a pair of feet' can be found in S. G. Mivart (1867), On the appendicular skeleton of the primates, *Philosophical Transactions of the Royal Society of London*, 157: 231, 370. The quote 'the lower kinds … types of structure' is from S. G. Mivart (1873), *Man and apes, an exposition of structural resemblances and differences bearing upon questions of affinity and origin*, Robert Hardwicke, London, p. 5, while 'a tangled web …natural selection' is from p. 176.

10. LINES IN THE SEA

Voltaire describes the apple tree and gravity story in Voltaire (1738), *Elements of Sir Isaac Newton's Philosophy*, trans John Hanna, Stephen Austen, London. The quote from Henry Bates, 'make for ourselves … the origin of species', is from H. W. Bates (1863), *The naturalist on the River Amazon, a record of adventures, habits of animals, sketches of Brazilian and Indian life, and aspects of nature under the equator, during eleven years of travel*, vol. 1, John Murray, London, quoted in S. Knapp (1999), *Footsteps in the forest: Alfred Russel Wallace in the Amazon*, Natural History Museum, London, p. 11, which also provided much of the information on Wallace's early life. For further biographical detail on Wallace see P. van Oosterzee (1997), *Where worlds collide: the Wallace line*, Reed Books, Melbourne; and J. L. Brooks (1984), *Just before the origin: Alfred Russel Wallace's theory of evolution*, Columbia University Press, New York.

Further detail on the work of Thomas Malthus can be found in T. R. Malthus (1803), *An essay on the principle of population; or, A view of its past and present effects on human happiness; with an inquiry into our prospects respecting the future removal or mitigation of the evils which it occasions*, J. Johnson & T. Bensley, London.

For a general introduction to the discovery of evolution and the nature of biological science before and after see D. Young (1992), *The discovery of evolution*, Cambridge University Press, Cambridge. Darwin and Wallace first outlined their theory of natural selection in C. Darwin and A. R. Wallace (1858), On the tendency of species to form varieties; and on the perpetuation of varieties by natural means of selection, *Journal of the Proceedings of the Linnean Society (Zoology)*, 3: 45–62. Wallace elaborated further in A. R. Wallace (1857), On the law which has regulated the introduction of new species, *Annals and Magazine of Natural History*, series 2, 14: 184–96. The quotes 'there suddenly flashed upon me…which managed to survive' and 'In this way…the origins of species' are from A. R. Wallace (1905), *My life: a record of events and opinion*, Bell, London, p. 363. The contributions of French scientists to early biogeographic research can be found in G. Buffon (1761), *Histoire naturelle, générale et particulière*, vol. 5, Imprimeries royale, Paris; and A. Candolle (1820), Essai elementaire de géographie botanique, in *Dictionnaire des sciences naturelles*, vol. 18, Flevrault, Strasbourg. Charles Lyell's influential work is C. Lyell (1830), *Principles of geology*, 3 vols, London. Many of Wallace's biogeographic ideas are outlined in A. R. Wallace (1859), The geographical distribution of birds, *Ibis*, 1: 449–54; A. R. Wallace (1860), On the zoological geography of the Malay Archipelago, *Journal of the Proceedings of the Linnean Society (Zoology)*, 2: 1104–8; and A. R. Wallace (1913), *The Malay Archipelago*, Macmillan, London (see p. 117 for the quote 'In so well-cultivated... natural history'; p. 120 for 'Birds were plentiful ... traveler's journey eastward'; and p. 123 for 'One small room... described were impossible').

Details on alternative lines are from P. L. Sclater (1858), On the general geographical distribution of the Class Aves, *Journal of the Proceedings of the Linnean Society (Zoology)*, 2: 130–45; A. Murray (1866), *The geographical distribution of mammals*, Day & Son, London; R. Lydekker (1896), *A geographical history of mammals*, Cambridge University Press, Cambridge; W. L. Sclater and P. L. Sclater (1899), *The geography of mammals*, Kegan, Paul, Trench & Trübner, London; and M. Weber (1902), *Der Indo-australische Archipel und die Gestichte seiner Teirwelt*, 46 pp., Jena, cited in Mayr (1944). Modern debates on the value of these and Wallace's lines can be found in E. Mayr (1944), Wallace's line in the light of recent zoogeographic studies, *Quarterly Review of Biology*, 19: 1–14; G. G. Simpson (1977), Too many lines; the limits of the Oriental and Australian zoogeographic regions, *Proceedings of the American Philosophical Society*, 121: 107–20; and J. A. Keast (1983), In the steps of Alfred Russel Wallace: biogeography of the Asian–Australian interchange zone, in R. W. Sims, J. H. Prince and P. E. S. Whalley (eds), *Evolution, time and space: the emergence of the biosphere*, Academic Press, London.

For information on the distribution of bird families derived from the fossil record see P. V. Rich and E. M. Thompson (1982), *The Fossil vertebrate record of Australia*, Monash University, Melbourne. For distribution as determined on the basis of molecular data see C. G. Sibley and J. E. Alquist (1985), The phylogeny and classification of the Australo-Papuan passerine fauna, *Emu*, 85: 1–14; and L. Christidis and R. Schodde (1991), Relationships of Australo-Papuan songbirds: protein evidence, *Ibis*, 133: 277–85.

Original data on Wallace specimens were derived from Rory O'Brien at Museum Victoria and the Natural History Museum, London (1871), Acquisitions Register (Ornithology), pp. 8–411 (Image capture files AV02008–AV02411). Full details of the analysis and interpretation of this data can be found in D. Clode and R. O'Brien (2001), Why Wallace drew the line: a re-analysis of Wallace's bird collections in the Malay Archipelago and the origins of biogeography, in I. Metcalfe, J. M. B. Smith, M. Morwood and I. Davidson (eds), *Faunal and floral migrations and evolution in SE Asia–Australasia*, A. A. Balkema Press, Lisse.

11. SHIFTING CONTINENTS

Details of the Spelaeogriphacea can be found in G. C. B. Poore and W. F. Humphries (1998), First record of Spelaeogriphacea from Australasia: a new genus and species from an aquifer in the arid Pilbara of Western Australia, *Crustaceana*, 71(7): 721–42.

Lyell's influential geological theories are published in C. Lyell (1830), *Principles of Geology*, 3 vols, London. Bishop Ussher's calculation of the earth's age is from J. Ussher (1650), *Annales Veteris Testamenti, a prima Muni Origine Deducti*, London. Cuvier's theory of catastrophism and extinction is outlined in G. Cuvier (1813), *Essay on the theory of the Earth*, trans. R. Kerr, William Blackwood, Edinburgh. Hutton's uniformitarian geological theory is outlined in J. Hutton (1795), *Theory of the earth, with proofs and illustrations*, 2 vols, Edinburgh. The theory of natural selection was first outlined in C. Darwin and A. R. Wallace (1858), On the tendency of species to form varieties; and on the perpetuation of varieties by natural means of selection, *Journal of the Proceedings of the Linnean Society (Zoology)*, 3: 45–62. Lamarck's transformationist evolutionary model was first published in J. P. B. Lamarck (1809), *Philosophie zoologique*, Dentu, Paris, but later popularised to an English audience in C. Lyell (1830), *Principles of geology*, 3 vols, London. The dispute between Cuvier and Saint-Hilaire is discussed in T. A. Appel (1987), *The Cuvier–Geoffroy debate: French biology in the decades before Darwin*, Oxford University Press, Oxford. The quote 'the belief that … of short duration' is from C. Darwin (1859), *On the origin of species by means of natural selection*, John Murray, London, p. 481.

The quote 'the great recklessness … afraid to wet its feet' is from A. P. Coleman (1916), Dry land in geology, *Bulletin of the Geological Society of America*, 27: 173. Wegener outlines his theory of continental drift in A. L. Wegener (1921), *Die Entstehung der Kontinente und Ozeane*, trans. J.G.A. Skerl (1924), as *The origin of continents and oceans*, Methuen & Co., London. For an overview of the history of continental drift from a southern hemisphere perspective see H. E. Le Grand (1988), *Drifting continents and shifting theories*, Cambridge University Press, Cambridge (whence is derived the quote 'most of earth's … quicker to see them'); and H. E. Le Grand (1991), Theories of the earth as seen from below, in R. W. Home and S. G. Kohlstedt (eds), *International science and national scientific identity*, Kluwer Academic Publishers, Netherlands.

Modern scientific debates on the origin of ratites are derived from J. Cracraft (1974), Phylogeny and origin of the ratite birds, *Ibis*, 116: 494–521; and A. Cooper et al. (2001), Complete mitochondrial genome sequences of two extinct moas clarify ratite evolution, *Nature*, 409 (6821): 704–7. The bowerbird research is from L. Christidis, P. R. Leeton and M. Westerman (1996), Were bowerbirds part of the New Zealand fauna? *Proceedings of the National Academy of Science of the United States of America*, 93(9): 3898–901.

Material on Macquarie Island invertebrates is from R. Marchant and P. Lillywhite (1994), Survey of stream invertebrate communities on Macquarie Island, *Australian Journal of Marine and Freshwater Research*, 45: 471–81; and T. O'Hara (1998), Origin of Macquarie Island echinoderms, *Polar Biology*, 20: 143–51.

12. LIFE ON MARS

Metrodorus is quoted 'To consider the Earth … grain will grow' by Plutarch (46–120), Philosophical essays: sentiments concerning nature with which philosophers were delighted, ch. V, Whether the universe is one single thing, published in *The Complete Works*, vol. 3: *Essays and Miscellanies* (1909), Crowell, New York.

General discussions on the history and possibility of life on Mars can be found in P. Davies (1998), *The fifth miracle: the search for the origin of life*, Allen Lane, Ringwood, Vic.; E. K. Gibson et al. (1999), *Planetary dreams: the quest to discover life beyond earth*, John Wiley & Sons, New York; and C. S. Romanek (1996), The case for relic life on Mars, *Scientific American*, August. Lowell outlined his theories of Martian life in P. Lowell (1895), *Mars*, Houghton Mifflin, Boston.

Hahn's observations of possible bacterial life on meteorites is from O. Hahn (1880), *Die Meteorite und ihre Organismen*, Tübingen, pp. 1–56. Fred Hoyle outlined his theory of the origin of life in space in F. Hoyle and N. C. Wickramasinghe (1981), *Space travellers: the bringers of life*, University of Cardiff Press, Cardiff; and F. Hoyle and N. C. Wickramasinghe (1981), *The origin of life*, University of Cardiff Press, Cardiff. Research into organic compounds in meteorites includes G. Claus and B. Nagy (1961), A microbiological examination of some Carbonaceous chondrites, *Nature*, 192: 594; B. Nagy, D. J. Hennessy and W. G. Meinschein (1961), Mass spectroscopic analysis of the Orgueil meteorite: evidence for biogenic hydrocarbons, *Annals of the New York Academy of Sciences*, 93: 25; and

B. Nagy (1975), *Carbonaceous meteorites*, Elsevier, Amsterdam. Further research and reviews have been completed by H. C. Urey (1966), Biological material in meteorites: a review, *Science*, 151: 157; F. Anders, R. Hayatsu and M. H. Studier (1973), Organic compounds in meteorites, *Science*, 182: 4114; and D. A. J. Seargent (1991), *Genesis stone? The Murchison meteorite and the beginnings of life*, Karagi Publications, The Entrance, NSW. Information on deep rock bacterial life forms can be found in J. K. Frederickson and T. C. Onstott (1996), Microbes deep inside the Earth, *Scientific American*, October.

President Clinton's announcement of the Martian meteorite discovery is documented in W. Clinton (1996), President Clinton's statement regarding Mars meteorite discovery, The White House, Office of the Press Secretary, 7 August. The original scientific paper was D. S. McKay et al. (1996), Search for past life on Mars: Possible relic biogenic activity in Martian meteorite ALH 84001, *Science*, 273: 924–30. More recent research on organic compounds in meteorites can be found in P. Ehrenfreund et al. (2001), Extraterrestrial amino acids in Orgueil and Ivuna: tracing the parent body of CI type carbonaceous chondrites, *Proceedings of the National Academy of Sciences USA*, 98(5): 2138–41; P. Ehrenfreund and S. B. Charnley (2000), Organic molecules in the interstellar medium, comets, and meteorites: a voyage from dark clouds to the early earth, *Annual Review of Astronomy and Astrophysics*, 38: 427–83; and R. B. Hoover (2004), Evidence for the detection of a fossilized cyanobacterial mat in a freshly fractured, interior surface of the Orgueil carbonaceous meteorite, paper presented at the *Instruments, Methods, and Missions for Astrobiology VIII Conference*, Denver. Further developments from the study of Comet Tempel 1 are in M. F. A'Hearn et al. (2005), Deep Impact: Excavating Comet Tempel 1, *Science*, online, 8 Sept 2005.

REFERENCES

---

1 CURIOUS COLLECTIONS

Arnold, K. R. (1992), Cabinets for the curious: practising science in early modern English museums, PhD thesis, Princeton University.

Bennet, T. (2004), *Pasts beyond memory: evolution, museums and colonialism*, Routledge, London.

Belk, R. H. (1995), *Collecting in a consumer society*, Routledge, New York.

Bonyhady, T. (2000), *The colonial earth*, Miegunyah Press, Melbourne.

Hooper-Greenhill, E. (1992), *Museums and the shaping of knowledge*, Routledge, London.

Kohlstedt, S. G. (1983), Australian museums of natural history: public priorities and scientific initiatives in the nineteenth century, *Historical Records of Australian Science*, 5 (4): 1–29.

Pescott, R. T. M. (1954), *Collections of a century: the history of the first hundred years of the National Museum of Victoria*, National Museum of Victoria, Melbourne.

Rasmussen, C. (2001), *A museum for the people: A history of Museum Victoria and its predecessors, 1854–2000*, Scribe Publications, Melbourne.

Sheets-Pyenson, S. (1988), *Cathedrals of science: the development of colonial natural history museums during the late nineteenth century*, McGill–Queens University Press, Montreal.

Dobzhansky, T. (1973), Nothing in biology makes sense except in the light of evolution, *American Biology Teacher*, 35: 125–9.

2 A BEAST NAMED SU

Banks, J. (1770), quoted in Parkins (1997).

Bennett, G. (1860), *Gatherings of a naturalist in Australasia*, John van Voorst, London.

de Bruijn, C. (1720), *Voyage to the Levant and travels into Moscovy, Persia, and the East Indies*, quoted in Younger (1988).

Calaby, J. H., Mack, G., and Ride, W. D. L. (1962), The application of the generic name Macropus Shaw 1790 and of other names commonly referred to the Grey Kangaroo. *Memoirs of the Queensland Museum* 14 (2): 25–31.

Cook, J. (1770), quoted in Parkins (1997).

Dampier, W. (1703), *A voyage to New Holland etc in the year 1699*, London, James Knapton (quoted in George, 1999).

Dixon, J., and Huxley, L. (1985), *Donald Thomson's mammals and fishes of Northern Australia*, Nelson, Melbourne.

Dixon, R. M. W., Ramson, W. S., and Thomas, M. (1990), *Australian Aboriginal words in English: their origin and meaning*, Oxford University Press, Melbourne.

Drake-Brockman, H. (1963), *Voyage to disaster: the life of Francisco Pelsaert*, Angus & Robertson, Sydney.

Flannery, T. F. (1995), *Mammals of the South-West Pacific and Molluccan Islands*, Reed Books, Chatswood, NSW.

Frith, H. J., and Calaby, J. H. (1969), *Kangaroos*, F. W. Cheshire, Melbourne.

George, A. S. (1999), *William Dampier in New Holland: Australia's first natural historian*, Bloomings Books, Melbourne.

Hartman, C. (1952), *Possums*, University of Texas, Austin.

Jacobs, H. T. T. M. (1971), *A treatise on the Moluccas (c. 1544)* Jesuit Historical Institute, Rome.

McIntyre, K. G. (1982), *The secret discovery of Australia: Portuguese ventures 250 years before Captain Cook*, Picador; Sydney (rev. and abridged edn).

Parkins, R. (1997), *H.M. Bark Endeavour: her place in Australian history*, Melbourne University Press, Melbourne.

Pelsaert, F. (1647) *The voyage of the Batavia*; trans. W. Siebenhaar (1897); republished (1994), Hordern House, Sydney.

Playford, P. (1998), *Voyage of discovery to Terra Australis, by William de Vlamingh, 1696–97*, Western Australian Museum, Perth.

Smith, B., and Wheeler, A. (1988) (eds), *The art of the First Fleet and other early Australian drawings*, Oxford University Press, Melbourne.

Strahan, R. (1990), *The Mammals of Australia*, Reed Books, Chatswood, NSW; republished (1995), Australian Museum, Sydney.

Torst, M. quoted in Playford (1998), p. 62.

Tyson, E. (1698), *The anatomy of an opossum: dissected at Gresham College*, Royal Society, London.

Whitley, G. P. (1970), *Early history of Australian zoology*, Royal Zoological Society of New South Wales, Sydney.

Witson, N. (1705), *Noord en Ost Tartarye*, Tweede Druk VI, Francois Halma, Amsterdam, pp. 179–83, quoted in Playford (1998), p. 28.

Younger, R. M. (1988), *Kangaroo: images through the ages*, Hutchinson Australia, Melbourne.

3 LOCAL KNOWLEDGE

Dunmore, J. (2002), *Jeanne Baret, first woman around the world (1766–1768)*, Auckland, Heritage Press.

Blandowski, W. (1857), Recent discoveries in Natural History on the Lower Murray, *Philosophical Institute of Victoria* 2 (1): 124–37.

Bowden, K. M. (1952), *George Bass, 1771–1803: his discoveries, romantic life and tragic disappearance*, Oxford University Press, Melbourne.

Clode, D. (2001), The respectful invader, in Rasmussen (2001), pp. 18–20.

—— (2005), Popular and professional communicators: Edith Coleman and Norman Wakefield, *Leaves from Our History*, Field Naturalists Club of Victoria 125th Anniversary Symposium.

Dixon, J. M., and Huxley, L. (1985), *Donald Thomson's mammals and fishes of northern Australia*, Nelson, Melbourne.

Dixon, R. M. W., Ramson, W. S., and Thomas, M. (1990), *Australian Aboriginal words in English: their origin and meaning*, Oxford University Press, Melbourne.

Eliot, G. (1871), *Middlemarch*, republished 1985 Penguin Classics, Aylesbury.

Gould, J. (1863), *Mammals of Australia*, J. Gould, London.

Mansergh, I., and Hercus, L. A. (1981), An Aboriginal vocabulary of the fauna of Gippsland, *Memoirs of the National Museum of Victoria*, 42: 107–22.

Pianka, E. R. (1986), *Ecology and natural history of desert lizards: analyses of the ecological niche and community structure*, Princeton University Press, Princeton.

Reid, J. R. W., Kerle, J. A., and Morton, S. R. (1993), Uluru fauna: The distribution and abundance of vertebrate fauna of Uluru (Ayers Rock–Mount Olga) National Park, NT, *Kowari*, 4.

Selzer, A. (2002), *Governors' wives in colonial Australia*, National Library of Australia, Canberra.

Spencer, W. B. (ed.) (1896), *Report on the work of the Horn Scientific Expedition to Central Australia: Part 1.—Introduction, Narrative and summary of results, supplement to zoological report, map*, Melville, Mullen & Slade, Melbourne.

—— (ed.) (1896), *Report on the work of the Horn Scientific Expedition to Central Australia: Part 2.—Zoology*, Melville, Mullen & Slade, Melbourne.

Stanbury, P. (1975), *100 years of Australian scientific explorations*, Holt, Rinehart & Winston, Sydney.

Sumner, R. (1993), *A woman in the wilderness: the story of Amalie Dietrich in Australia*, University of New South Wales Press, Sydney.

Tunbridge, D. (1991), *The story of the Flinders Ranges mammals*, Kangaroo Press, Kenthurst, NSW.

Waddy, J. A. (1988), *Classification of plants and animals from a Groote Eylandt Aboriginal point of view*, North Australia Research Unit, Australian National University, Canberra.

Wilson, E. O. (1992), *The diversity of life*, Penguin, London.

Yen, A. L., Gillen, J., Gillespie, R., Vanderwal, R., and the Mutitjulu Community, (1997), A preliminary assessment of Anangu knowledge of central Australian invertebrates, *Memoirs of the Museum of Victoria*, 546(2): 631–4.

4 WATER, WATER EVERYWHERE

Cannon, M. (1991), *Old Melbourne Town: before the gold rush*, Loch Haven, Melbourne.

Dingle, T., and Rasmussen, C. (1991), *Vital connections: Melbourne and its Board of Works, 1891–1991*, McPhee Gribble, Melbourne.

Dunn, G. M., and Connor, D. J. (1991), *Management of transpiration loss and water yield in Mountain Ash: a final report*, Faculty of Agriculture and Forestry, University of Melbourne, Melbourne.

Ellison, K. (2003), Investing in nature, *The Source* (Melbourne Water), 26.

Lee, R. (1980), *Forest Hydrology*, Columbia University Press, New York.

Marchant, R. (1987), *Thomson River post-construction monitoring program: changes in the benthic invertebrate communities of the Thomson River, Victoria, after dam construction*, A report to the Department of Water Resources, Victoria.

—— (1988), Changes in the benthic invertebrate communities of the Thomson River, Southeastern Australia, after dam construction, *Regulated Rivers: Research and Management*, 4: 7189.

Marchant, R., and Hehir, G. (2002), The use of AUSRIVAS predictive models to assess the response of lotic macroinvertebrates to dams in south-east Australia, *Freshwater Biology*, 47: 1003–50.

McClellan, P. (1998), *Sydney Water Inquiry: final report*, NSW Premier's Department, Sydney.

Mueller, F. (1871), *Forest culture in relation to industrial pursuits*, Government Printer, Melbourne.

NSW Department of Health (1998), The Sydney Water Incident: July–September 1998, *NSW Public Health Bulletin*; 9: 91–4.

Pollio, M. V. (27 BCE), *On architecture*, Book VIII, ch. 1; trans. J. Gwilt

(1826), *The Architecture of Marcus Vitruvius Pollio*, Priestley & Weale, London.

5 FORESTS OF FIRE

A'Hearn, M. F., et al. (2005), Deep impact: excavating Comet Tempel 1, *Science*, online, 8 Sept 2005.

Attiwill, P. M. (1994), The disturbance of forest ecosystems: the ecological basis for conservation management, *Forest Ecology and Management*, 63: 247–300.

—— (1994), Ecological disturbance and the conservative management of eucalypt forests in Australia, *Forest Ecology and Management*, 63: 301–46.

Brown, G. W., and Nelson, J. L. (1992), *Habitat utilisation by heliothermic reptiles of different successional stages of* Eucalyptus regnans *(Mountain Ash) forest in the central highlands*, VSP Technical Report No. 17, Department of Conservation and Natural Resources, Melbourne.

Brown, G. W., et al. (1989), *Flora and Fauna of the Acheron Forest Block, Central Highlands, Victoria*, Department of Conservation, Forests and Lands, Melbourne.

Clode, D., and Burgman, M. (eds) (1997), Cultural values of old-growth forests, in *Joint old-growth forest project: summary report*, NSW Parks and Wildlife Service and NSW State Forests, Sydney.

Darwin, C. (1845), *Journal of researches into the natural history and geology of the countries visited during the voyage of the HMS 'Beagle' round the world*, Ward, Lock & Co., London.

Dumont d'Urville, J. (1830–35), *Voyage de la corvette l'Astrolabe, exécuté par ordre du Roi pendant les années 1826, 1827, 1828, 1829, sous le commandement de M. Jules S-C Dumont d'Urville*, 17 vols, Tastu & Cie, Paris.

Griffiths, T. (1992), *Secrets of the forest: discovering history in Melbourne's Ash range*, Allen and Unwin, St Leonards, NSW.

Howard, T. M. (1975), Litter fauna in *Nothofagus cunninghamii* forests, *Proceedings of the Royal Society of Victoria* 87: 207–13.

Lawson, H. (1896), His country after all, *While the billy boils*, Angus & Robertson, Sydney.

Lindenmayer, D. (1996), *Wildlife and woodchips: Leadbeater's Possum, a test case for sustainable forestry*, University of NSW Press, Sydney.

Pyne, S. J. (1991), Antipodal fire: bushfire research in Australia and America, in R. W. Home and S. G. Kohlstedt (eds), *International Science and National Scientific Identity*, Kluwer, Boston, pp. 225–87.

Rosenman, H. (1988–92), *Two voyages to the South Seas by Captain Jules S-C Dumont d'Urville*, Melbourne University Press, Melbourne.

6 THE MYSTERY OF THE REAPPEARING POSSUMS

Brazenor, C. W. (1932), A re-examination of *Gymnobelideus leadbeateri* McCoy, *Australian Zoologist*, 7 (2): 106–9.

—— (1946), Last chapter to come. A history of Victoria's rarest possum, *Wild Life*, 8: 383–4.

Broom, R. (1895), On a small fossil Marsupial with large grooved premolars. *Proceedings of the Linnaean Society of New South Wales*, series 2, 10(4): 563–7.

—— (1895), On a small fossil Petaurus-like Marsupial, *Proceedings of the Linnaean Society of New South Wales*, series 2, 10(4): 568–70.

Calaby, J., Dimpel, H., and McTaggart Cowan, I. (1971), *The Mountain Pigmy Possum* Burramys parvus *Broom (Marsupialia) in the Kosciusko National Park, New South Wales*, Division of Wildlife Research Technical Paper No. 23, CSIRO, Australia.

Clode, D. (2001), The elusive Leadbeater, in Rasmussen (2001), pp. 253–5.

—— (2002), Norman Wakefield, in J. Ritchie and D. Langmore (eds), *Australian Dictionary of Biography*, vol. 16, Melbourne University Press, Melbourne.

Dixon, J. M. (1971), *Burramys parvus* Broom (Marsupialia) from Falls Creek area of the Bogong High Plains, Victoria, *Victorian Naturalist*, 88: 133–8.

Lindenmayer, D. (1996), *Wildlife and woodchips: Leadbeater's Possum, a test case for sustainable forestry*, University of NSW Press, Sydney.

Mansergh, I., and Broome, L. (1994), *The Mountain Pygmy Possum of the Australian Alps*, University of New South Wales Press, Sydney.

McCoy, F. (1867), On a new genus of phalanger, *Annals of the Magazine of Natural History*, 3: 287–8.

Osborne, M. J., et al. (2000), Genetic distinctness of isolated populations of an endangered marsupial, the Mountain Pygmy-possum, *Burramys parvus*, *Molecular Ecology*, 9 (5): 609–13.

Ride, W. D. L. (1956), The affinities of *Burramys parvus* Broom, a fossil marsupial, *Proceedings of the Zoological Society of London*, 127 (3): 413–29.

—— (1970), *A guide to native mammals of Australia*, Oxford University Press, Oxford.

Rolls, E. (1984), *They all ran wild: the animals and plants that plague Australia*, Angus & Robertson, Sydney.

Smith, A. P. (1980), The diet and ecology of Leadbeater's Possum and the Sugar Glider, PhD thesis, Monash University, Melbourne.

Spencer, W. B. (1899), Remarks on a rare marsupial, *Victorian Naturalist*, 16 (7): 105–6.

Strahan, R. (1990), *The Mammals of Australia*, Reed Books, Chatswood, NSW; republished (1995), Australian Museum, Sydney.

van Dyck, S. (1993), The taxonomy and distribution of *Petaurus gracilis* (Marsupialia: Petauridae), with notes on its ecology and conservation status, *Memoirs of the Queensland Museum*, 33: 77–122.

Wilkinson, H. E. (1961), The rediscovery of Leadbeater's Possum *Gymnobelideus leadbeateri* McCoy, *Victorian Naturalist*, 78: 97–102.

7 THE CASE OF THE MISSING MOLLUSC

Agassiz, L. (1840), Mémoire sur les trigonies, in *Etudes critiques sur les mollusques fossiles*, Petitpierre, Neuchatel, quoted in Gould (1993), p. 448.

Darragh, T. A. (1986), The Cainozoic trigoniidae of Australia, *Alcheringa*, 10: 1–34.

Darwin, C. (1859), *On the origin of species by means of natural selection*, John Murray, London.

Dumont d'Urville, J. (1830–35), *Voyage de la corvette l'Astrolabe, exécuté par ordre du Roi pendant les années 1826, 1827, 1828, 1829, sous le commandement de M. Jules S-C Dumont D'Urville*, 17 vols, Tastu & Cie, Paris.

Gould, S. J. (1993), A foot soldier for evolution, *Eight little piggies: reflections in natural history*, Norton, New York.

Hall, T. S. (1901), Growth stages in modern Trigonias belonging to the section Pectinatae, *Proceedings of the Royal Society of Victoria*, 14: 17–21.

Jenkins, H. M. (1865), On the occurrence of a Tertiary species of trigonia in Australia, *Quarterly Journal of Science*, 2 (April): 362–4.

—— (1866), On the occurrence of a Recent species of Trigonia (*T. lamarckii*) in Tertiary deposits in Australia, *Geological Magazine*, 3: 201–3.

Lamarck, J. P. B. (1804), Sur une nouvelle espèce de Trigonie et sur une nouvelle espèce d'Huître, découverte dans le voyage du capitaine Baudin, *Annales du Muséum d'Histoire Naturelle*, 4: 353–4.

—— (1809), *Philosophie zoologique*, Dentu, Paris.

Lyell, C. (1830), *Principles of geology*, 3 vols, London.

McCoy, F. (1870), *Lecture on the order and plan of creation, delivered before the Early Closing Association, Melbourne, 1869–70*, pp. 23, 25, 26–7, 31–2, National Library of Australia.

—— (1875), On a third new Tertiary species of trigonia, *Annals and Magazine of Natural History*, series 4, 15: 316–17.

Mozely-Moyal, A. (1976), *Scientists in nineteenth century Australia: a documentary history*, Cassell Australia, Melbourne.

Péron, F. (1809), *A voyage of discovery to the southern hemisphere, performed by order of the Emperor Napoleon during the years 1801, 1802, 1803 and*

1804, vol. 1, Richard Phillips, London; republished (1975), Marsh Walsh, Melbourne.

Rosenman, H. (1988–92), *Two voyages to the South Seas by Captain Jules S-C Dumont d'Urville*, Melbourne University Press, Melbourne.

8 BRAINBOX

Barton, R. A., Purvis, A., and Harvey, P. H. (1995), Evolutionary radiation of visual and olfactory brain systems in primates, bats and insectivores, *Philosophical Transactions of the Royal Society of London B*, 348: 381–92.

Butler, A. B., and Hodds, W. (1996), *Comparative vertebrate neuroanatomy: Evolution and adaptation*, Wiley–Liss, New York.

Buffetaut, E. (2004), Polar dinosaurs and the question of dinosaur extinction: a brief review, *Palaeogeography, Palaeoclimatology, Palaeoecology*, 214 (3): 225–31.

Constantine, A., et al. (1998), Periglacial environments and polar dinosaurs, *South African Journal of Science*, 94: 137–41.

Galton, P. M. (1974), The Ornithiscian dinosaur Hypsilophodon from the Wealden of the Isle of Wight, *Bulletin of the British Museum (Natural History) Geology*, 25: 1–149.

Hopson, J. A. (1977), Relative brain size and behavior in archosaurian reptiles, *Annual Review of Ecology and Systematics*, 8: 429–48.

Huber, R., et al. (1997), Microhabitat use, trophic patterns and the evolution of brain structure in African cichlids, *Brain Behavior and Evolution*, 50 (3): 167–82.

Huxley, T. H. (1869), On Hypsilophodont a new genus of Dinosauria, *Abstracts of the Proceedings of the Geological Society of London*, 204: 3–4.

Potter, H. D. (1969), Structural characteristics of cell and fiber populations in the optic tectum of the frog (*Rana catesbeiana*). *Journal of Comparative Neurology* 136: 203–32.

Rich, P. V., et al. (1988), Evidence for low temperatures and biologic diversity in Cretaceous high latitudes of Australia, *Science*, 242: 1403–6.

Rich, T. H., and Rich, P. V. (1989), Polar dinosaurs and biotas of the early cretaceous of southeastern Australia, *National Geographic Research*, 5 (1): 15–53.

Rich, T. H., Vickers-Rich, P., and Gangloff, R. A. (2002), Polar dinosaurs, *Science*, 295: 979–80.

Roth, J. J., and Roth, E. C. (1980), The parietal-pineal complex among paleovertebrates: evidence for temperature regulation, in R. D. K. Thomas and E. C. Olson (eds), *A cold look at warm-blooded dinosaurs*, West View Press, British Columbia, pp. 189–231.

Spencer, W. B. (1885), *On the presence and structure of the pineal eye in Lacertilia*, J. E. Adlard, London.

Ulinski, P. S., Dacey, D. M., and Sereno, M. I. (1992), Optic tectum, in C. Gans and P. S. Ulinski (eds), *Biology of the reptilia*, vol. 17: *Neurology C*, pp. 241–366.

Vickers-Rich, P., and Rich, T. H. (1993), Australia's Polar Dinosaurs, *Scientific American*, July: 40–5.

9 THE APE CASE

Butcher, B. W. (1988), Gorilla warfare in Melbourne: Halford, Huxley and 'Man's place in Nature', in R. W. Home (ed.), *Australian science in the making*, Cambridge University Press, Cambridge, pp. 153–69.

Darwin, C. (1859), *On the origin of species by means of natural selection*, John Murray, London.

—— (1875), *The descent of man*, John Murray, London.

Halford, G. B. (1863), *Not like man bimanous and biped, nor yet quadromanous, but cheiropodous*, Wilson & Mackinnon, Melbourne.

—— (1864), *Lines of demarcation between man, gorilla and macaque*, Wilson & Mackinnon, Melbourne.

Huxley, T. H. (1863), *Evidence as to man's place in nature*, Williams & Norgate, London.

Lucas, J. R. (1979), Wilberforce and Huxley: a legendary encounter, *Historical Journal*, 22 (2): 313–30.

McCoy, F. (1863), Letter to Prof. Halford, 13th July 1863, *National Museum Outward Letter Book*, vol. 2, Sept 1861–May 1865, p. 217.

—— (1865), Gorillas at the National Museum, *Argus*, 20 June, p. 5.

Mivart, S. G. (1867), On the appendicular skeleton of the primates, *Philosophical Transactions of the Royal Society of London*, 157: 299–429.

—— (1873), *Man and apes, an exposition of structural resemblances and differences bearing upon questions of affinity and origin*, Robert Hardwicke, London.

Owen, R. (1861), The gorilla and the negro, *The Athenaeum*, 23 March: 395–6.

Smith, C. U. M. (1997), Worlds in collision: Owen and Huxley on the brain, *Science in Context*, 10(2): 343–65.

Spencer, H. (1851), *Social statics, or, The conditions essential to human happiness specified, and the first of them developed*, John Chapman, London, vol. VI, ch. 2.

Tyson, E. (1699), *Orang-outang, sive* Homo Sylvestris*: or, the Anatomy of a pygmie*; facs. edn (1966), Dawsons of Pall Mall, London.

Wallace, A. R. (1857), On the law which has regulated the introduction of new species, *Annals and Magazine of Natural History*, series 2, 14: 184–96.

10 LINES IN THE SEA

Bates, H. W. (1863), *The naturalist on the River Amazon, a record of adventures, habits of animals, sketches of Brazilian and Indian life, and aspects of nature under the equator, during eleven years of travel*, vol. 1, John Murray, London.

Buffon, G. (1761), *Histoire naturelle, générale et particulière*, vol. 5, Imprimeries royale, Paris.

Candolle, A. (1820), Essai Elementaire de Geographie Botanique, *Dictionnaire des sciences naturelles*, vol. 18, Flevrault, Strasbourg.

Christidis, L., and Schodde, R. (1991), Relationships of Australo-Papuan songbirds—protein evidence, *Ibis*, 133: 277–85.

Clode, D., and O'Brien, R. (2001), Why Wallace drew the line: a re-analysis of Wallace's bird collections in the Malay Archipelago and the origins of biogeography, in I. Metcalfe, J. M. B. Smith, M. Morwood and I. Davidson (eds), *Faunal and floral migrations and evolution in SE Asia-Australasia*, A. A. Balkema Press, Lisse.

Darwin, C., and Wallace, A. R. (1858), On the tendency of species to form varieties; and on the perpetuation of varieties by natural means of selection, *Journal of the Proceedings of the Linnean Society (Zoology)*, 3: 45–62.

Keast, J. A. (1983), In the steps of Alfred Russel Wallace: biogeography of the Asian–Australian Interchange Zone, in R. W. Sims, J. H. Prince and P. E. S. Whalley (eds), *Evolution, time and space: the emergence of the biosphere*, Academic Press, London.

Knapp, S. (1999), *Footsteps in the forest: Alfred Russel Wallace in the Amazon*, Natural History Museum, London.

Lydekker, R. (1896), *A geographical history of mammals*, Cambridge University Press, Cambridge.

Lyell, C. (1830), *Principles of geology*, 3 vols, London.

Malthus, T. R. (1803), *An essay on the principle of population; or, A view of its past and present effects on human happiness; with an inquiry into our prospects respecting the future removal or mitigation of the evils which it occasions*, J. Johnson & T. Bensley, London.

Mayr, E. (1944), Wallace's line in the light of recent zoogeographic studies, *Quarterly Review of Biology*, 19: 1–14.

Murray, A. (1866), *The geographical distribution of mammals*, Day & Son, London.

Natural History Museum, London (1871), Acquisitions Register (Ornithology), pp. 8–411 (Image capture files AV02008–411).

Rich, P. V., and Thompson, E. M. (1982), *The fossil vertebrate record of Australia*, Monash University, Melbourne.

Sclater, P. L. (1858), On the general geographical distribution of the class Aves, *Journal of the Proceedings of the Linnean Society (Zoology)*, 2: 130–45.

Sclater, W. L., and Sclater P. L. (1899), *The geography of mammals*, Kegan, Paul, Trench & Trübner, London.

Sibley, C. G., and Alquist, J. E. (1985), The phylogeny and classification of the Australo-Papuan passerine fauna, *Emu*, 85: 1–14.

Simpson G. G. (1977), Too many lines; the limits of the Oriental and Australian zoogeographic regions, *Proceedings of the American Philosophical Society*, 121: 107–20.

van Oosterzee, P. (1997), *Where worlds collide: the Wallace line*, Reed Books, Melbourne.

Wallace, A. R. (1857), On the law which has regulated the introduction of new species, *Annals and Magazine of Natural History*, series 2, 14: 184–96.

—— (1859), The geographical distribution of birds, *Ibis*, 1: 449–54.

—— (1860), On the zoological geography of the Malay Archipelago, *Journal of the Proceedings of the Linnean Society (Zoology)*, 2: 1104–8.

—— (1905) *My life: a record of events and opinion*, Bell, London.

—— (1913), *The Malay Archipelago*, Macmillan & Co., London.

Weber, M. (1902), *Der Indo-australische Archipel und die Gestichte seiner Teirwelt*. 46 pp., Jena, cited in Mayr (1944).

Young, D. (1992), *The discovery of evolution*, Cambridge University Press, Cambridge.

11 SHIFTING CONTINENTS

Appel, T. A. (1987), *The Cuvier–Geoffroy debate: French biology in the decades before Darwin*, Oxford University Press: Oxford.

Christidis, L., Leeton, P. R., and Westerman, M. (1996), Were bowerbirds part of the New Zealand fauna? *Proceedings of the National Academy of Science of the United States of America*, 93(9): 3898–901.

Coleman, A. P. (1916), Dry land in geology, *Bulletin of the Geological Society of America*, 27: 171–204.

Cooper, A., et al. (2001), Complete mitochondrial genome sequences of two extinct moas clarify ratite evolution. *Nature*, 409 (6821): 704–7.

Cracraft, J. (1974), Phylogeny and origin of the ratite birds, *Ibis*, 116: 494–521.

Cuvier, Georges. (1813), *Essay on the theory of the Earth*, trans. R. Kerr, William Blackwood, Edinburgh.

Darwin, C. (1859), *On the origin of species by means of natural selection*, John Murray, London.

Darwin, C., and Wallace, A. R. (1858), On the tendency of species to form varieties; and on the perpetuation of varieties by natural means of selection, *Journal of the Proceedings of the Linnean Society (Zoology)*, 3: 45–62.

Hutton, J. (1795), *Theory of the earth, with proofs and illustrations*, 2 vols, Edinburgh.

Lamarck, J. P. B. (1809), *Philosophie zoologique*, Dentu, Paris.

Le Grand, H. E. (1988), *Drifting continents and shifting theories*, Cambridge University Press, Cambridge.

—— (1991), Theories of the earth as seen from below, in R. W. Home and S. G. Kohlstedt (eds), *International science and national scientific identity*, Kluwer Academic Publishers, Netherlands.

Lyell, C. (1830), *Principles of geology*, 3 vols, London.

Marchant, R., and Lillywhite, P. (1994), Survey of stream invertebrate communities on Macquarie Island, *Australian Journal of Marine and Freshwater Research*, 45: 471–81.

O'Hara, T. (1998), Origin of Macquarie Island echinoderms, *Polar Biology*, 20: 143–51.

Poore, G. C. B., and Humphries, W. F. (1998), First record of Spelaeogriphacea from Australasia: a new genus and species from an aquifer in the arid Pilbara of Western Australia, *Crustaceana*, 71 (7): 721–42.

Ussher, J (1650), *Annales Veteris testamenti, a prima muni origine deducti*, London.

Wegener. A. L. (1921), *Die Entstehung der Kontinente und Ozeane*; trans. J. G. A. Skerl (1924), *The origin of continents and oceans*, Methuen & Co., London.

12 LIFE ON MARS

Ahern, L. D., and Yen, A. L. (1977), A comparison of the invertebrate fauna under Eucalyptus and Pinus forests in the Otway Ranges, Victoria, *Proceedings of the Royal Society of Victoria*, 89: 127–36.

Anders, F., Hayatsu, R., and Studier, M. H. (1973), Organic compounds in meteorites, *Science*, 182: 4114.

Claus, G., and Nagy, B. (1961), A microbiological examination of some carbonaceous chondrites, *Nature*, 192: 594.

Clinton, W. (1996), President Clinton's statement regarding Mars meteorite discovery, The White House, Office of the Press Secretary, 7 August.

Davies, P. (1998), *The fifth miracle: the search for the origin of life*, Ringwood, Vic., Allen Lane.

Ehrenfreund, P., and Charnley, S. B. (2000), Organic molecules in the interstellar medium, comets, and meteorites: a voyage from dark clouds to the early earth, *Annual Review of Astronomy and Astrophysics* 38: 427–83.

Ehrenfreund, P., et al. (2001), Extraterrestrial amino acids in Orgueil and Ivuna: tracing the parent body of CI type carbonaceous chondrites, *Proceedings of the National Academy of Sciences USA*, 98(5), 2138–41.

Frederickson, J. K., and Onstott, T. C. (1996), Microbes deep inside the Earth, *Scientific American*, October.

Gibson, E. K., et al. (1996), The case for relic life on Mars, *Scientific American*, August.

Hahn, O. (1880), *Die Meteorite und ihre Organismen*, Tübingen, pp. 1–56.

Hoover, R. B. (2004), Evidence for the detection of a fossilized cyanobacterial mat in a freshly fractured interior surface of the Orgueil carbonaceous meteorite, paper presented at the *Instruments, Methods, and Missions for Astrobiology VIII Conference*, Denver.

Hoyle, F., and Wickramasinghe, N. C. (1981), *Space travellers: the bringers of life*, University of Cardiff Press, Cardiff.

—— and —— (1981), *The Origin of Life*, University of Cardiff Press, Cardiff.

Lowell, P. (1895), *Mars*, Houghton Mifflin, Boston.

McKay, D. S., et al. (1996), Search for past life on Mars: Possible relic biogenic activity in Martian meteorite ALH 84001, *Science*, 273: 924–30.

Nagy, B. (1975), *Carbonaceous meteorites*, Elsevier, Amsterdam.

Nagy, B., Hennessy, D. J., and Meinschein, W. G. (1961), Mass spectroscopic analysis of the Orgueil meteorite: evidence for biogenic hydrocarbons, *Annals of the New York Academy of Sciences*, 93: 25.

Plutarch (1909), *The Complete Works*, vol. 3: *Essays and Miscellanies*, Crowell, New York.

Seargent, D. A. J. (1991), *Genesis stone? The Murchison meteorite and the beginnings of life*, Karagi Publications, The Entrance, NSW.

Shapiro, R. (1999), *Planetary dreams: the quest to discover life beyond earth*, John Wiley & Sons, New York.

Urey, H. C. (1966), Biological material in meteorites: a review, *Science*, 151: 157.

ILLUSTRATIONS

PLATES

FIGURES

INDEX

Page numbers in italics refer to illustrations, plates and their accompanying captions.